Mathematical Modelling for Engineering and Physical Applications

This edited volume from mathematical modelling experts employs a structured approach to showcase the latest research and provide a comprehensive guide to the principles, techniques and practical applications of mathematical modelling in the fields of engineering and the physical sciences.

Readers will gain an insight into the fundamentals of mathematical modelling, its application in engineering and the physical sciences, recent advances in the field, real-world case studies and challenges and future directions in the field. The book first introduces readers to the fundamentals of mathematical modelling, including the definition, scope and steps involved in the modelling process. Mathematical tools and techniques relevant to modelling are explained. The book then addresses the application of mathematical modelling in engineering and the physical sciences. It covers various domains, such as mechanical, biomedical, chemical, computer and environmental engineering, as well as green energy, physics and other physical science applications. Furthermore, the book explores the integration of data-driven approaches, leveraging the power of big data and analytics to enhance modelling capabilities. Through real-life case studies, readers gain practical insights into how mathematical modelling is applied in various industries and scientific disciplines. These case studies analyse the modelling process, challenges faced and outcomes achieved. Readers will gain a solid foundation to develop effective mathematical models and apply them to solve complex engineering and physical problems. This foundation will also enable them to tackle problems and make informed decisions in their own projects. Finally, readers will be able to anticipate and adapt to evolving technologies and stay ahead in their respective fields.

This book is a valuable resource primarily for undergraduate and graduate students seeking a comprehensive guide to this important discipline. Researchers, engineers and scientists will also benefit from the discussions within. Overall, the book empowers readers with the knowledge, skills and practical insights to effectively apply mathematical modelling in engineering and the physical sciences.

Mathematical Modelling for Engineering and Physical Applications

Practical Examples and Case Studies

Edited by Mohd Hafiz Mohd, Noor Atinah Ahmad, Syakila Ahmad, Md Yushalify Misro and Mohamad Hekarl Uzir

CRC Press
Taylor & Francis Group
Boca Raton London New York

CRC Press is an imprint of the
Taylor & Francis Group, an **informa** business

Designed cover image: Internet of Things (IOT) technology with AR (Augmented Reality) on VR dashboard: Shutterstock

First edition published 2025
by CRC Press
2385 NW Executive Center Drive, Suite 320, Boca Raton FL 33431

and by CRC Press
4 Park Square, Milton Park, Abingdon, Oxon, OX14 4RN

CRC Press is an imprint of Taylor & Francis Group, LLC

ISBN: 978-1-032-67823-8 (hbk)
ISBN: 978-1-032-67856-6 (pbk)
ISBN: 978-1-032-67857-3 (ebk)

DOI: 10.1201/9781032678573

Typeset in Minion
by Apex CoVantage, LLC

Contents

Editors

Mohd Hafiz Mohd earned a BSc (Hons) in mathematics at the Imperial College London, an MSc in mathematics at the Universiti Sains Malaysia and a PhD in computational and applied mathematics at the University of Canterbury, New Zealand. He is an Associate Professor at Universiti Sains Malaysia, researching mathematical modelling in biology, engineering and distinct complex systems.

Noor Atinah Ahmad earned a BSc (Hons) in mathematics at the University of Bristol, UK, and a PhD in mathematics at the University of Southampton, UK. She is an Associate Professor at the Universiti Sains Malaysia, and her research interests include data-driven modelling of dynamical systems.

Syakila Ahmad is a Senior Lecturer at the School of Mathematical Sciences, Universiti Sains Malaysia. She earned a PhD in fluid dynamics at the Universiti Putra Malaysia. Her research interests include fluid dynamics, convective heat and mass transfer and mathematical modelling.

Md Yushalify Misro earned a BSc (Hons) and an MSc (Hons) in mathematics and a PhD in computer-aided graphic design at the Universiti Sains Malaysia. He joined the faculty of the Universiti Sains Malaysia in 2018 and has since researched various areas, including trigonometric and fractional Bézier curves and surfaces, developable surfaces, numerical splines and traffic simulation.

Mohamad Hekarl Uzir earned a BEng (Hons) in chemical engineering at the University of Leeds, an MSc in advanced chemical engineering at the University of London and a PhD in biochemical engineering at the University College London. He is an Associate Professor at the Universiti Sains Malaysia, specialising in biocatalysis, biotransformation and modelling of biocatalytic and bioprocess systems.

Contributors

Salaudeen Abdulwaheed Adebayo is an academic staff member of Sa'adu Zungur University, formerly Bauch State University, Nigeria. He earned a BSc in mathematics at Bayero University, Kano, Nigeria, and a MSc at the Universiti Sains Malaysia, and he is currently pursuing a PhD in the School of Mathematical Sciences at the Universiti Sains Malaysia. His areas of interest are computational mathematics, applied mathematics, neural networks and mathematical modelling. He has been actively involved in research studies contributing to the understanding of mathematics.

Afifurrahman joined the Universitas Islam Negeri Mataram, Indonesia, as a mathematics lecturer in 2015. He teaches mathematics at the undergraduate level, conducts research in applied mathematics and is involved in community services. He is also an active member of the Indonesian Mathematical Society (IndoMS), a nonprofit professional organisation. He serves as head of the Research Department for the Bali and Nusa Tenggara regions. His research interests include neurodynamics and its application to system biology. Through mathematical modelling and numerical simulation, he investigates the behavior of large bodies of interacting neurons. He uses the theories from dynamical systems and statistical mechanics to understand the underlying mechanisms of collective dynamics that emerge in neural network models.

Majid Khan Majahar Ali is a senior lecturer at the School of Mathematical Sciences, Universiti Sains Malaysia. He earned a PhD in statistical modelling at Universiti Malaysia Sabah. His research interests include statistical mathematics, big data, machine learning, operational research and mathematical modelling.

Mohammad Arafat is a PhD candidate in mechanical engineering technology at Universiti Tun Hussein Onn, Malaysia. He earned a BSc in industrial engineering at Telkom University, Indonesia. His research is focused on computational fluid dynamics with a particular focus on flow mechanisms and noise generation in aerospace and aeronautical systems through mathematical modelling and simulation techniques.

Zainudin Arsad earned a BSc in actuarial mathematics and statistics, a PhD in stochastic investment models at Heriot-Watt University in Edinburgh and an MSc in statistics at the University of Sheffield. His area of research is time series analysis and econometric modelling, with a main interest in the application of the Kalman Filter estimation technique to various financial and economic topics. He is also interested in tourism and environmental research, including determinants of tourism demand, attributes of tourists' satisfaction levels and investigation of global warming. In addition to research and teaching, he conducts workshops and short courses on various econometric modelling and statistical analysing techniques.

Norhashidah Awang is a senior lecturer at the School of Mathematical Sciences, Universiti Sains Malaysia. She earned a BSc (Hons) in mathematics at the Universiti Sains Malaysia; an MSc in industrial and systems engineering at the National University of Singapore, specialising in operations research; and a PhD at the Universiti Putra Malaysia. Her research interests include statistical modelling, with a particular emphasis on time series analysis, spatial analysis and the statistical modelling of environmental data.

'Afifi Md Desa is a lecturer at the Institute of Engineering Mathematics, Universiti Malaysia Perlis (UniMAP), and is currently pursuing a PhD at the School of Mathematical Sciences, Universiti Sains Malaysia. His research focuses on mathematical modelling of dynamical systems, utilising continuation and simulation methods.

Sabastine Emmanuel began his academic journey in 2011 as a lecturer at the Federal University Lokoja, Kogi State, Nigeria. He is in his second year of the PhD program at the School of Mathematical Sciences, Universiti Sains Malaysia, specialising in computational mathematics. His research is centered on neuro-numerical analysis, numerical methods, neural networks and mathematical modelling, with a focus on solving complex real-world challenges using these advanced techniques.

Ahmad Fudholi is an associate professor at the Solar Energy Research Institute, Research Complex, Universiti Kebangsaan Malaysia. He is also Director for Energy Research in the Indonesia National Research and Innovation Agency. His research interests are solar energy, renewable energy, carbon emission and techno-economic analysis.

R. U. Gobithaasan earned a PhD in mathematics at the Universiti Sains Malaysia and in 2009 joined the Faculty of Computer Science and Mathematics, Universiti Malaysia Terengganu (UMT) as a lecturer. He was promoted to associate professor in 2015, and in 2023 he returned to the Universiti Sains Malaysia's School of Mathematical Sciences to teaching and research in computational mathematics, focusing on topological data analysis, geometric modelling and graphs. His research applies these methods to time series, climate and image datasets. Over the years he has successfully led projects supported by national and international agencies.

Hirohide Haga is a professor of computer science at Doshisha University, Kyoto, Japan. He earned a BEng and an MEng in electronics at Doshisha University and a PhD in computer science at Kyoto University in 1978, 1980 and 1995, respectively. Before joining Doshisha, he was a research engineer at Hitachi Ltd, one of the largest leading engineering companies in Japan. His research interests include programming language, multi-agent simulation and painting analysis (image processing). He is also interested in theoretical computer science fields, including combinatorial theory and computability theory.

Nuralia Nisa Mohd Nor Hisham earned a BSc in applied statistics in 2021 and an MSc in statistics in 2022 at the Universiti Sains Malaysia. Since 2023 she has been a manufacturing systems engineer at a semiconductor company, leveraging the programming skills she acquired during her academic journey.

Izuan Amin Ishak is a senior lecturer in the Department of Mechanical Engineering Technology, Universiti Tun Hussein Onn, Malaysia. He earned a PhD in mechanical engineering at Universiti Teknologi Malaysia. His research interests include computational fluid dynamics, vehicle aerodynamics and turbulence flow, with a strong focus on mathematical modelling in engineering applications.

Nurul Syuhada Ismail is a lecturer at the Centre for Pre-University Studies, Universiti Malaysia Sarawak (UNIMAS), Sarawak, Malaysia. She earned a B.App.Sc. (Hons) in mathematical modelling and a MSc in mathematics at the Universiti Sains Malaysia and a PhD in fluid dynamics at the Universiti Putra Malaysia (UPM). Her research interests include applied mathematics and fluid dynamics problems with stability analysis, boundary layer flow problem, heat transfer and mathematical modelling.

Fengli Lan is a researcher at the Muscle Corporation. His research interests include manipulator control, path generation and programming.

Kenjiro T. Miura is a specially appointed professor and professor emeritus at the Graduate School of Science, Shizuoka University. His research interests include geometric modelling, especially aesthetic curve and surface design and its applications to CAD/CAM systems.

Sora Nakao joined the Department of Intelligent Information Engineering and Science at Doshisha University in 2013 and graduated in 2016. While at Doshisha, he studied information processing. His bachelor's thesis is about multi-agent simulation and its application to the numerical analysis of linear simultaneous equations. After finishing his study at Doshisha, he joined TIS, a Japanese software development company in Osaka.

Wei Shan Ng is pursuing a PhD at the School of Mathematical Sciences, Universiti Sains Malaysia. Her research focuses on the intersection of online learning, multi-output prediction and dynamic mode decomposition.

Wen Eng Ong is a senior lecturer in the School of Mathematical Sciences, Universiti Sains Malaysia. Her research interests include the reconstruction of curves and surfaces, shortest path algorithms and applications, computational mathematics and machine learning.

Amirah Rahman is a senior lecturer in the School of Mathematical Sciences at the Universiti Sains Malaysia. Her research interests include discrete optimisation, metaheuristics and other areas related to optimisation and operations research applied to real-world problems.

Norshafira Ramli is a senior lecturer at the School of Mathematical Sciences, Universiti Sains Malaysia. She graduated from the Universiti

Sains Malaysia with a PhD in mathematical modelling. Her research interests are applied mathematics that include fluid mechanics, boundary layer flow and heat transfer and mathematical modelling.

Saratha Sathasivam earned a BSc (Ed) and an MSc at the Universiti Sains Malaysia and a PhD at the Universiti Malaya, Malaysia. She is an associate professor in the School of Mathematical Sciences, Universiti Sains Malaysia. She has published widely in journals and proceedings and has collaborated with researchers from different countries. Her research interests include neural networks, mathematical modelling, numerical and neuro-numerical analysis, agent-based modelling, data mining and constrained optimisation problems.

Veekeong Saw earned an MSc in mathematics at the Universiti Sains Malaysia. He now works on research and development projects related to path planning and navigation in industrial automation. His research interests include issues related to grid-based navigation, multi-agent path planning, resource allocation and scheduling problems and multicriteria optimisation problems.

Raja Aqib Shamim is a PhD student in the School of Mathematical Sciences, Universiti Sains Malaysia. His research interests include statistical mathematics, big data, machine learning, operations research and mathematical modelling.

Hirofumi Tamai is the President of Muscle Corporation. His research interests include motor control, mechanical design and robotics.

Dan Wang is a postdoctoral fellow in the Department of Electronic, Electrical and Mechanical Engineering, University of Shanghai for Science and Technology. Her research interests include computer-aided design, especially volume parametric modelling and its applications to CAD/CAE systems.

Nurul Nasyrah Zaini is an undergraduate student at the School of Mathematical Sciences, Universiti Sains Malaysia. She completed her studies in 2022. Her research interests are mathematical modelling and numerical methods.

CHAPTER 1

Introduction to Mathematical Modelling for Engineering and Physical Applications

Mohd Hafiz Mohd, Noor Atinah Ahmad, Syakila Ahmad, Md Yushalify Misro and Mohamad Hekarl Uzir

1.1 INTRODUCTORY REMARKS

The book *Mathematical Modelling for Engineering and Physical Applications* addresses the growing need for an additional resource that integrates mathematical modelling techniques with their practical applications in engineering and the physical sciences. As is widely recognised, engineering and physical sciences have become increasingly reliant on mathematical modelling techniques to solve complex problems, design reliable systems and optimise related real-life processes. Therefore, this book offers a detailed understanding of the principles and techniques required to effectively modelling engineering and physical problems under consideration.

One of the objectives of this book is to showcase the latest research in mathematical modelling for engineering and physical sciences, which will

DOI: 10.1201/9781032678573-1

serve as a valuable text for scientific researchers, students and professionals. This book also aims to provide a guide to the principles, techniques and applications of mathematical modelling in the fields of engineering and the physical sciences. It employs a systematic approach to deliver the information and covers a range of topics, including the fundamentals of mathematical modelling, its application in engineering and physical sciences, recent advances in the field and real-world case studies. Each chapter begins by introducing readers to the fundamentals of selected mathematical modelling problems, including the definition, scope and steps involved in examining the modelling systems. Then, mathematical tools and techniques relevant to modelling are explained further, ensuring that readers have a solid foundation and prerequisite knowledge to better understand the real-life modelling examples and case studies that will be presented later.

1.2 OVERVIEW OF CASE STUDIES

In particular, the subsequent chapters discuss the applications of mathematical modelling in engineering—including financial, biomedical, chemical, computer, environmental and transport engineering—as well as green energy, nanofluid and other physical science fields. Readers will embark on a journey through the diverse landscape of mathematical modelling, exploring how various mathematical tools and techniques are employed to address complex challenges across engineering and physical sciences. From the unpredictable nature of neuronal dynamics and stock prices, which are examined through stochastic modelling perspectives, to the deterministic modelling approaches in biochemical reactions and fluid dynamics problems, readers will see how different models cater to the specific needs of each engineering and physical application.

Each modelling aspect in the respective chapters is also structured for readers to better understand the use of data-driven and statistical approaches in environmental and multi-agent systems, highlighting their importance in simulating complex behaviours in natural ecosystems. Some chapters and case studies also touch on the cutting-edge applications in transportation and robotics where computer-aided geometric design and operations research techniques play a crucial role in enhancing the efficiency and performance of these physical systems.

1.2.1 Stochastic Modelling

In Chapters 2 and 3, stock price and neuronal dynamics are examined using stochastic approaches. In neuronal systems, stochastic modelling techniques are essential for understanding the complex and high variability in brain function, contributing to phenomena such as neuronal firing patterns and signal propagation. Additionally, stock prices are influenced by a wide range of factors, including supply and demand dynamics, company performance, geopolitical events, inflation and market sentiment. For this reason, stochastic models are often employed extensively to capture the randomness and uncertainty inherent in stock price movements.

1.2.2 Deterministic Modelling

Distinct deterministic models that integrate fluid engineering and enzyme catalysis are discussed at length in several chapters, particularly Chapters 4 through 6. The analysis does not only focus on the modelling technique, but also examines the stability of system by applying the dynamical systems method. Some simulation results from the analysis could later be used, for instance, to design a reactor control system in a process plant. In the field of fluid engineering, computational fluid dynamics is the method used by engineers to monitor distinct engineering systems. Modelling and simulation of such systems play major roles in predicting and monitoring flow in industrial pipelines, enhancing the aerodynamics of train systems and allowing for realistic simulations of heat transfer and fluid dynamics at the nanoscale.

1.2.3 Multi-Agent Systems and Data-Driven Modelling

A multi-agent system (MAS) is a computational system composed of multiple autonomous agents that interact with each other and their environment. Each agent in a MAS is typically capable of independent action, possesses some degree of autonomy and may have specific goals or objectives. These agents can perceive their environment, make decisions and communicate with other agents to achieve individual or collective goals. In Chapter 7, an application of MAS in the numerical simulation of linear simultaneous equations is described. The MAS-based solution is realised using particle swarm optimisation (PSO). Unlike ordinary PSO, where particles are placed uniformly random in the search space, the method in Chapter 7 uses Cauchy distribution-based particle placement.

In the study of the dynamics of complex systems, the primary focus of MAS is on simulating the interactions between autonomous agents. On the other hand, data-driven modelling involves building models based on observed data to capture underlying patterns, make predictions or derive insights. Instead of starting with a predefined mathematical or theoretical model, data-driven modelling iteratively refines models based on data insights. Models can be adjusted, validated and improved as more data becomes available or as the understanding of the system evolves. Techniques like machine learning, statistical modelling and time series analysis fall into this category. In Chapter 8, a real-time data-driven identification of nonlinear dynamical systems approach is used to track abrupt changes in the system dynamics. This modelling framework is based on a recursive regression that updates model parameters incrementally with incoming data. Abrupt changes in the system parameters are identified by tracking the sudden change in the average local separation of orbits in the phase space of the dynamical system, informed by the data. A comparative study of machine learning and statistical modelling is conducted in Chapter 9 that involves the forecasting of particulate matter concentrations in Penang, Malaysia. The machine learning method used in the study is random forest, while statistical modelling is performed using the univariate autoregressive integrated moving average with generalised autoregressive conditional heteroscedasticity (ARIMA-GARCH). This study underpins the importance of a good predictive model to help policymakers make informed decisions about environmental regulations and interventions to mitigate pollution, thus enabling targeted efforts to improve air quality.

1.2.4 Computer-Aided Geometric Design and Operations Research

Chapters 10 and 11 focus on the basics of trajectories and path planning, using ideas from computer-aided geometric design and operations research techniques. Chapter 10 introduces the κ-curve, explaining its properties and how this curve can be used in robot trajectory generation. The key points of the κ-curve can be adjusted dynamically, and the curve can be extended into a space curve by changing it at inflection points.

In Chapter 11, the discussion moves to path planning for unmanned ground vehicles. The problem is set up like a travelling salesman problem with unique start and end points. The proposed solution divides the checkpoints into clusters, solving the problem in two stages. A global 2-opt heuristic is also used to improve the solution. Tests show that the quality of the path depends greatly on how the checkpoints are grouped.

Additionally, Chapter 12 details the design, construction and testing of a solar drying system for red seaweed, exploring its drying kinetics and analysing energy and exergy efficiencies. Key performance metrics, including specific energy consumption and exergy efficiency, highlight the system's potential for renewable biofuel production.

These case studies are presented throughout the book to provide practical context and illustrate the application of mathematical modelling in diverse real-life scenarios. Interested readers are guided to gain insights into how these mathematical models are employed to solve practical problems and analyse selected engineering and physical systems. Special attention is given to various applications and case studies on the techniques of mathematical modelling in examining distinct engineering and physical science fields. The inclusion of case studies throughout the book bridges theory with real-world applications. It also showcases the practical implementation of mathematics in diverse scenarios, providing readers with valuable insights and illustrating the relevance of the discussed modelling concepts.

1.3 CLOSING REMARKS AND WAY FORWARD

As we move forward, it is essential to continue integrating these diverse modelling concepts, recognising that the complexity of real-world problems often requires a multidisciplinary perspective. Future work should focus on enhancing the synergy between deterministic, stochastic and data-driven models to create more robust, adaptive and predictive systems. Moreover, as computational power and data availability continue to grow, the development of hybrid models that can seamlessly combine the strengths of various methodologies will be the key to advancing both theoretical understanding and practical applications. By fostering collaboration across disciplines and continually refining our modelling frameworks with emerging data and computational techniques, we can ensure that mathematical modelling remains at the forefront of innovation in engineering and physical sciences.

CHAPTER 2

Forecasting Stock Price Fluctuation Using Geometric Brownian Motion Modelling

Sabastine Emmanuel, Saratha Sathasivam, Salaudeen Abdulwaheed Adebayo and Nurul Nasyrah Zaini

2.1 INTRODUCTION

Forecasting stock prices is a highly challenging task, primarily because it is profoundly influenced by numerous factors, such as government policy, economic conditions, unforeseen incidents and the financial health of the underlying companies and businesses (Fama, 1970). It poses difficulties due to its dynamic, ever-changing and rapidly changing nature (Shahi et al., 2020). Characterised by notable volatility and non-linearity, predicted stock prices often deviate slightly from what is obtainable in reality. However, the pursuit of developing numerical and technical approaches for forecasting stock prices should not be underestimated or undervalued. Stock price prediction has long been an interesting topic that has been thoroughly investigated by researchers and scholars from various fields of study (Mensah et al., 2023). Jabbarzadeh et al. (2016) emphasised the

DOI: 10.1201/9781032678573-2

importance of continuous improvement of predictive models as they remain essential tools, even if their performance is not entirely precise. Precise and accurate price prediction models are of immense value to all stakeholders. Nevertheless, the inherent volatility and intricacy of stock markets render precise forecasts elusive and arduous. Any attempt to glean insight into the trend in the dataset comes with a price, which revolves around getting a foolproof model that can capture the intricacy and nuance within the stock price.

Researchers and financial experts have developed a wide array of approaches and machine learning techniques that can handle prediction tasks effectively. Nemes and Butoi (2013) asserted that time series predictive models can forecast future stock prices or patterns based on historical data arranged in chronological order. Such models are built to capture patterns, relationships and underlying dynamics in order to make accurate predictions via exponential smoothing and regression. ARIMA is regarded as one of the best predictive models for stock price prediction in business and finance (Madziwa et al., 2022). However, when adequately modelled and responsive to factors that affect stock price changes, Nemes and Butoi (2013) highlighted the effectiveness and reliability of data-driven models in prediction. Another effective predictive model is the autoregressive distributed LAG (ARDL) model, which was reported as one of the most superb techniques for capturing stock price fluctuation in both the long and short terms. It is able to identify vital information in datasets when there are abnormal changes in the stock price.

Neural networks have gained massive popularity due to their potent strategies, resulting in an upsurge in predictive models. Their adaptability is tailored to specific tasks, and their adeptness in recognising intricate nonlinear affinities and patterns within data enhances their allure for stock price forecasting (Zahedi & Rounaghi, 2015). When employing an artificial neural network as a predictive model, a historical dataset is collected alongside pertinent attributes encompassing prior prices, volume metrics and other pertinent factors; potentially a well-trained network can then be deployed to make predictions (Mahdiani & Khamehchi, 2016). One area of study that has earned a good reputation is using geometric Brownian motion (GMB) to model stock prices, which began in 1900 (Brătian et al., 2022; Delgado, 2019); since its appearance, it has undergone a lot of modification, improvement and applicability (Black & Scholes, 1973; Djeutcha et al., 2019). The following are some of the works that have been modified and used GBM to predict the price of stocks. Campbell et al. (1997) reported that the success of GBM as a predictive model revolves around the effective modelling of drift and volatility terms. These two parameters enable the modelling of stock price movements under the assumption that stock prices follow a genuine random walk pattern, so that the stock price for today would be the best indicator of tomorrow's stock price. Dash (2019) reported that the applicability of GBM presumes that stock returns within consecutive and uniform time intervals should exhibit both independence and a normal distribution, and concluded that the random walk hypothesis can be seen as an outcome derived from the application of GBM. Several models have been proposed, including the Black-Scholes model, the Merton model and the Heston model, among others, which have recorded massive success in the prediction of price stock. In this chapter, we will deploy the GBM and solve it with some notable numerical methods to predict the fluctuation in the stock price within the years under review to guide individuals towards investing their fortune in Malayan Banking Berhad (Maybank), a bank that has been reputed to be one of the top five banks in the ASEAN region for over 10 years. Maybank has consistently reported a steady rise in net interest income, growing from RM4.5 billion in 2007 to RM12.2 billion in 2017, reflecting its continuous development and stability.

2.2 DATA AND METHODOLOGY

Historical stock price data for Maybank for the years 2011–2022 was collected (Malayan Banking Berhad Maybank Stock Historical Prices & Data, Yahoo Finance). The data were prepared to suit our model by conversion of the daily data to yearly data using averaging techniques, and we equally calculated some important parameters needed to carry out this study.

From Figure 2.1, we extracted data from Maybank stock prices for the years 2011 to 2022 and tabulated it as shown in Table 2.1.

The information in Table 2.1 is used to calculate the mean growth rate (MGR), which is the average at which a stock price increases or decreases over a specified period. This information is particularly useful in financial analysis to comprehend the average yearly increase or decrease in stock prices. As such, this idea is rooted in compound interest or growth. By calculating the annual growth rate for each year, we are essentially

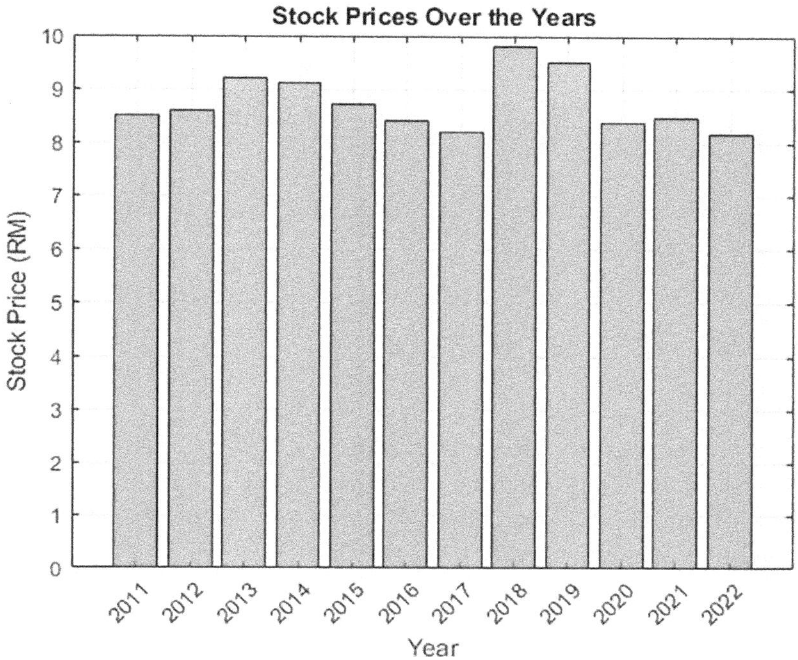

FIGURE 2.1 Maybank stock price for 2011 to 2022.

TABLE 2.1 Maybank Stock Prices for 2011 to 2022

Year	**2011**	**2012**	**2013**	**2014**	**2015**	**2016**	**2017**	**2018**	**2019**	**2020**	**2021**	**2022**
Stock price (RM)	8.50	8.58	9.20	9.10	8.72	8.40	8.20	9.80	9.50	8.37	8.46	8.16

quantifying the percentage change from one year to the next. By averaging these individual growth rates, we approximate the average growth rate for the entire duration. It is widely employed in time series data, especially in financial contexts, where the goal is to gauge the overall trend in a quantity across long time intervals. It offers a straightforward means to summarise fluctuations and grasp the typical behaviour of the quantity over time.

Calculate the annual growth rate for each year:

$$Annual\,Growth\,Rate = \left(\frac{Stock\ price\ in\ \ current\ year}{Stock\ price\ in\ \ previous\ year}\right) - 1 \tag{2.1}$$

Using the relation in Eq. (2.1) and Table 2.1, we arrived at the data in Table 2.2.

TABLE 2.2 Annual Growth Rate for Each Year

Year	2011	2012	2013	2014	2015	2016	2017	2018	2019	2020	2021	2022
Growth Rate (%)	0	0.94	7.21	−1.10	−4.10	−3.67	−2.38	19.51	−3.06	−11.94	1.28	−3.49

Using the stock exchange returns in Table 2.2, we calculated the MGR using the formula in Eq. (2.2).

$$MGR = \left(\prod_{i+1}^{n} (1 + R_i) \right)^{\frac{1}{n}} - 1 \tag{2.2}$$

where n is the number of years (in our case, it is 11 years) and R_i are growth rates. We designated the MGR by μ and it has a numerical value of $\mu = 1.3706$. The calculated growth rate of $\mu = 1.3706$ or $\mu = -1.3706$ resulted in unrealistic results since the data spikes and follows no specific pattern across the dataset. We opted for the values of μ that can model and produce a similar pattern to the real dataset, and we arrived at $\mu = 0.0035$ as one of many suitable values that produces minimal error and best describes the real stock price data under consideration. We equally considered the value $\mu = 0.0040$ for the same purpose, to ascertain the impact of differences of 0.0005 on our choice of μ.

2.3 MATHEMATICAL FORMULATION

GBM, which assumes constant volatility, is used as the underlying mathematical framework for stock price prediction. The choice of the GBM model for stock price prediction is attributed to its ability to capture the stochastic nature of stock prices and its widespread acceptance in finance. GBM's suitability arises from its ability to incorporate random fluctuations in stock prices (Farida et al., 2018), its alignment with the log-normal distribution observed in financial data and its continuous-time framework that mirrors the nature of trading in financial markets. The model's simplicity makes it computationally efficient and allows for an intuitive interpretation of its parameters, such as drift and volatility. GBM serves as a foundational model for stock price prediction, providing a benchmark for complex models (Brătian et al., 2022). Despite its strengths, GBM has limitations, including the assumption of constant volatility, and its inability to account for certain real-world factors might make it not be a good approximation (Al-Harthy, 2007).

2.4 GEOMETRIC BROWNIAN MOTION

GBM has been used as a suitable approach to approximate financial phenomena such as the evolution of stock prices over time, accounting for the influence of stochastic noise. This model can be expressed as follows:

$$\frac{dS}{S} = \mu dt + \sigma dX \tag{2.3}$$

where dS is the change in stock price over a small time interval dt, S is the current value of the asset, μ is the drift or expected growth rate of the asset per unit of time, σ is the volatility of the asset and represents the standard deviation of the asset's returns per unit of time and dX is a Wiener process that represents the random component, with dX following a normal distribution with mean 0 and variance dt and dX is independent of dt. This equation captures the continuous-time dynamics of an asset's price under the assumption that the asset's returns follow a log-normal distribution. It is a fundamental component of many financial models, including option pricing models like the Black-Scholes model.

Setting volatility, $\sigma = 0$, Eq. (2.3) reduces to

$$\frac{dS}{S} = \mu dt \tag{2.4}$$

Integrating both sides of Eq. (2.4), we have

$$\text{In} S = ut + C \tag{2.5}$$

with initial condition $(t_0) = S_0$, where S_0 is the value of the asset at the time $t = 0$, we solve for C and substitute its value into Eq. (2.5) and get

$$\text{In} S = ut + \text{In} S_0 - ut_0 \tag{2.6}$$

When we simplify and rearrange, we have

$$S(t) = S_0 e^{\mu(t-t_0)} \tag{2.7}$$

where $S(t)$ is the stock price at the time t, while S_0 and t_0 are initial values for stock price and time, respectively. Eq. (2.7) represents the exponential growth solution and can be used to obtain the value of stock at any particular time in the presence of a drift value.

2.5 MULTI-STEP METHOD

The multi-step method, also known as the linear multiple steps method (Konechnaya et al., 2021), is a numerical method for solving ordinary differential equations (ODEs). Its application complements the initial use of a single-step method to acquire the approximation solution of an ODE. A single-step approach refers to procedures that advance the solution from one point to the next without needing many sub-steps inside each interval period. The classical Euler numerical method is the most prominent type of single-step methodology. The multi-step method, unlike the single-step method, leverages solution approximations by improving accuracy and computation time. However, multi-steps are generally used when a high-quality solution is required. Many numerical methods for solving ODEs fall under this category. However, this chapter focuses specifically on evaluating the Adams-Bashforth predictor-corrector method, the three-step predictor-corrector (3-step P-C) method and the Runge-Kutta 4th-order method.

2.5.1 Adams-Bashforth Method

The Adams-Bashforth method, also known as the Adams-Bashforth-Moulton method, is made up of two sections: the predictor phase and the corrector phase. The predictor phase employs the explicit Adams-Bashforth method, while the corrector phase involves the implicit Adams-Moulton method. Because of their multi-step nature, both the Adams-Bashforth and the 3-step P-C methods necessitate an initial value for the ODE, y'. To be able to use the multi-step method, single-step methods must be utilised to get the few refined solutions of the given ODE. The predictor part computes roughly an approximation of the desired solution of the unknown function, while the corrector part corrects the approximation from the predictor part. The corrector can be applied more than once, depending on user choice and the nature of the problem under consideration. The general representation of a specific multistep method called the Adams-Bashforth method involves the integration of the form

$$y(t_{n+1}) = y(t_n) + \int_{t_n}^{t_{n+1}} \left(f(t, y(t))\right) dt. \tag{2.8}$$

where $y(t_{n+1})$ is the current solution, $y(t)$ is the previous solution, $f(t, y(t)) = p_4(x)$ is described in Eq. (2.9) and often called Lagrange polynomial. The derivation of the Adams-Bashforth method formula is obtained by interpolating the polynomial for $f(t, y(t))$ that passes through the four distinct points (t_{n-3}, f_{n-3}), (t_{n-2}, f_{n-2}), (t_{n-1}, f_{n-1}) and (t_n, f_n). Interpolating polynomials can be written as

$$p_4(x) = f_{n-3}L_{n-3}(t) + f_{n-2}L_{n-2}(t) + f_{n-1}L_{n-1}(t) + f_nL_n(t) \tag{2.9}$$

where $L_{n-3}(t), L_{n-2}(t), L_{n-1}(t)$ and $L_n(t)$ are called Lagrange polynomial basis. Substituting Eq. (2.9) into Eq. (2.8) and simplifying leads to

$$y_{n+1} = y_n + f_{n-3}\int_{t_n}^{t_{n+1}} L_{n-3}(t)dt + f_{n-2}\int_{t_n}^{t_{n+1}} L_{n-2}(t)dt$$
$$+ f_{n-1}\int_{t_n}^{t_{n+1}} L_{n-1}(t)dt + f_n\int_{t_n}^{t_{n+1}} L_n(t)dt \tag{2.10}$$

where f_{n-3}, f_{n-2}, f_{n-1} and f_n are function evaluations or function values of $f(t, y)$ evaluated at respected values of t_i. Solving Eq. (2.10) and substituting the limits of integration give Eq. (2.11), which is the predictor part of the Adams-Bashforth formula

$$S_{i+1} = S_i + \frac{h}{24}\left(55f_i - 59f_{i-1} + 37f_{i-2} - 9f_{i-3}\right) \tag{2.11}$$

where S_{i+1} and S_i are the current stock price and the previous stock price, respectively. However, due to the implicit nature of the Adams-Moulton technique, which involves solving nonlinear equations at each time step, it may not be practicable. Alternatively, we can create an Adams-Moulton predictor-corrector method by combining an Adams-Moulton method with an Adams-Bashforth method. Such a process works as follows: we compute a first approximation to y_{n+1} using the Adams-Bashforth method and designate t as $\hat{y}_{n+1}$, then evaluate $f(t_{n+1}, \hat{y}_{n+1})$ and then utilise the Adams-Moulton method as if it were an explicit method; we compute y_{n+1} but instead of solving the equation, we use $f(t_{n+1}, \hat{y}_{n+1})$ in place of $f(t_{n+1},)$, the corrector Adams-Moulton, is shown in Eq. (2.12).

$$S_{i+1} = S_i + \frac{h}{24}\left(9f_{i+1} + 19f_i + 5f_{i-1} - f_{i-2}\right) \tag{2.12}$$

2.5.2 Three-Step Predictor-Corrector (3-Step P-C) Formula

The 3-step P-C method is a numerical technique employed for solving ODEs. Its primary advantage lies in its higher accuracy compared to single-step methods. This method utilises three previous solutions to calculate the next solution, resulting in improved accuracy due to the incorporation of more information from past steps. By combining both predictor and corrector steps, it mitigates errors and enhances stability, making it suitable for complex and dynamic systems.

Moreover, the 3-step P-C method demonstrates good convergence properties, which are essential for long-term predictions and simulations. Its versatility makes it applicable across various fields, from physics and engineering to finance and economics. The method's balanced trade-off between accuracy and computational complexity makes it a valuable tool for obtaining precise numerical solutions to differential equations. We start by using a single step (Euler method) to approximate the solution at the prediction stage to obtain y_{n+1}^{p}, it is given as

$$y_{n+1}^{p} = y_n + hf(x_n, y_n). \tag{2.13}$$

The correction part is obtained, using

$$y_{n+1}^{c} = y_n + \frac{h}{2}\left[f(x_n, y_n) + \mathrm{f}\left(x_{n+1}, y_{n+1}^{p}\right)\right]. \tag{2.14}$$

This is updated to obtain the value of y_{n+1} as

$$y_{n+1} = \frac{h}{2}\left[f(x_n, y_n) + \mathrm{f}\left(x_{n+1}, y_{n+1}^{c}\right)\right]. \tag{2.15}$$

y_{n+1} is obtained as the stock price at the end of the financial year, for $n = 1, 2, 3, 4, \ldots$.

2.5.3 Runge-Kutta 4th-Order Method (RK4)

In the realm of mathematical techniques for solving ODEs, the Runge-Kutta 4th-order method emerges as a significant method, deriving its roots from the Taylor series. The Taylor series is a well-known numerical tool used in solving ODEs by approximating lower-order derivatives. Building upon this foundation, the RK4 method was formulated to enhance solution accuracy. The RK4 method unfolds in four distinct variations, each characterised by a different order. Notably, higher-order methods within the RK family yield more accurate results. As a result, it takes the spotlight as the most advanced choice among these RK techniques, offering a refined level of precision in solving differential equations. It is regarded as most the popular explicit numerical method for ODEs (Asif Arefin, 2020; Islam, 2015). It requires the initial value of the problem to be solved, after which one can apply it to obtain the numerical solution. RK4 is used to find the approximate value of an unknown function for some given values of dependent variables.

In general, given that $\frac{dy}{dx} = f(x, y)$ and $y(x_0) = y_0$.

Let $n = 0, 1, 2, \ldots \in Z$ and h as the step size, the formula to compute the solution of ODE using RK4 is given as

$$y_{n+1} = y_n + \frac{h}{6}(k_1 + 2k_2 + 2k_3 + k_4) \tag{2.19}$$

where

$$k_1 = f(t_i, y_i),$$

$$k_2 = f\left(t_i + \frac{h}{2}, y_i + \frac{hk_1}{2}\right),$$

$$k_3 = f\left(t_i + \frac{h}{2}, y_i + \frac{hk_2}{2}\right),$$

$$k_4 = f(t_i + h, y_i + hk_3).$$

where t, y denote the time in a year and growth rate in percentage, respectively; k_1 is the slope at the beginning of the interval; and k_2 denotes the slope at the midpoint of the interval, which is calculated using by moving half a step from the initial point in both time and growth rate. k_3 denotes another slope at the midpoint, but it uses the improved estimate k_2 to refine the value of growth rate, and k_4 is the slope at the end of the interval, which is calculated by moving a step-size from the initial point, both in time and growth rate. The approximated solution at each point can be computed iteratively using y_{n+1} at any specified time t.

2.6 PROCEDURE FOR IMPLEMENTATION OF THE NUMERICAL SIMULATION

To simulate the numerical methods for predicting stock prices, we used the following procedure:

|—Initialisation

| |—Set initial values

|—Loop until desired year is reached

| |—Euler's method (for A-B P-C and 3-step P-C)

| | |—Calculate next stock price using Euler method

| | |—Update time and stock price

| |—Prediction Phase (for A-B P-C and 3-step P-C)

| | |—Use previous values to predict next stock price

| |—Correction Phase (For A-B P-C and 3-step P-C)

| | |—Calculate corrected value using corrector formula

| |—Runge-Kutta 4th-order steps (For RK4)

| | |—Calculate k_1, k_2, k_3, and k_4 using RK4 formulas

| |—Update stock price and time

|—End loop

|—Calculate errors (optional)

|—Generate graphs and results

End.

2.7 PRESENTATION OF NUMERICAL RESULTS

Table 2.3 represents the simulated results obtained using the MATLAB programming language as numerical results of the Adams-Bashforth predictor-corrector, three-step predictor-corrector and Runge-Kutta of 4th order using $\mu = 0.0035$.

Table 2.4 depicts the numerical results of the Adams-Bashforth predictor-corrector, three-step predictor-corrector and Runge-Kutta 4th order using $\mu = 0.0040$.

Upon careful examination, it becomes evident that all the numerical techniques used yield consistent results. This consistency arises from the high effectiveness of these numerical methods in solving the ODE under consideration. Consequently, it can be inferred that they exhibit similar levels of error when compared to the actual data from the stock price. Tables 2.5 and 2.6 provide a detailed analysis of the errors associated with different values of μ.

TABLE 2.3 Stock Price Prediction for 2011 to 2022 Obtained Using $\mu = 0.0035$

		Predicted Stock Price (RM)		
Year (t)	**Real Stock Price**	**Runge-Kutta 4th Order**	**3-Step P-C**	**Adams-Bashforth P-C**
2011	8.50	8.5	8.5	8.5
2012	8.58	8.5298	8.5298	8.5298
2013	9.20	8.5597	8.5597	8.5597
2014	9.10	8.5897	8.5897	8.5897
2015	8.72	8.6198	8.6198	8.6198
2016	8.40	8.6501	8.6501	8.6501
2017	8.20	8.6804	8.6804	8.6804
2018	9.50	8.7108	8.7108	8.7108
2019	9.50	8.7414	8.7414	8.7414
2020	8.36	8.772	8.772	8.772
2021	8.46	8.8028	8.8028	8.8028
2022	8.16	8.8336	8.8336	8.8336

TABLE 2.4 Stock Price Prediction for 2011 to 2022 Obtained Using $\mu = 0.0040$

		Predicted Stock Price (RM)		
Year (t)	**Real Stock Price**	**Runge-Kutta 4th Order**	**3-Step P-C**	**Adams-Bashforth P-C**
2011	8.50	8.5	8.5	8.5
2012	8.58	8.5341	8.5341	8.5341
2013	9.20	8.5683	8.5683	8.5683
2014	9.10	8.6026	8.6026	8.6026
2015	8.72	8.6371	8.6371	8.6371
2016	8.40	8.6717	8.6717	8.6717
2017	8.20	8.7065	8.7065	8.7065
2018	9.50	8.7414	8.7414	8.7414
2019	9.50	8.7764	8.7764	8.7764
2020	8.36	8.8116	8.8116	8.8116
2021	8.46	8.8469	8.8469	8.8469
2022	8.16	8.8824	8.8824	8.8824

The graphical representations of the numerical simulation in Tables 2.5 and 2.6 are shown in Figures 2.2 and 2.3, respectively, to comprehend the dynamics of stock price predictions using numerical methods. By plotting the results from the Adams-Bashforth predictor-corrector (AB P-C), three-step predictor-corrector and Runge-Kutta of 4th order we can compare the accuracy and stability of each method over time. Graphs typically

TABLE 2.5 Recorded Error for Each of the Numerical Methods $\mu = 0.0035$

		Obtained Results			Absolute Error = |Exact Value − Approximated Value|			Relative Error = Absolute Error/Exact Value		
Year	Exact Value (RM)	Runge-Kutta 4th-Order Method	3-Step P-C Method	Adams-Bashforth Method	Runge-Kutta 4th-Order Method	3-Step P-C Method	Adams-Bashforth Method	Runge-Kutta 4th-Order Method	3-Step P-C Method	Adams-Bashforth Method
2011	8.50	8.5	8.5	8.5	0	0	0	0	0	0
2012	8.58	8.5298	8.5298	8.5298	0.0502	0.0502	0.0502	0.0058	0.0058	0.0058
2013	9.20	8.5597	8.5597	8.5597	0.6403	0.6403	0.6403	0.0695	0.0695	0.0695
2014	9.10	8.5897	8.5897	8.5897	0.5103	0.5103	0.5103	0.0560	0.0560	0.0560
2015	8.72	8.6198	8.6198	8.6198	0.1002	0.1002	0.1002	0.0114	0.0114	0.0114
2016	8.40	8.6501	8.6501	8.6501	0.2501	0.2501	0.2501	0.0297	0.0297	0.0297
2017	8.20	8.6804	8.6804	8.6804	0.4804	0.4804	0.4804	0.0585	0.0585	0.0585
2018	9.50	8.7108	8.7108	8.7108	1.0892	1.0892	1.0892	0.1111	0.1111	0.1111
2019	9.50	8.7414	8.7414	8.7414	0.7586	0.7586	0.7586	0.0798	0.0798	0.0798
2020	8.36	8.772	8.772	8.772	0.412	0.412	0.412	0.0492	0.0492	0.0492
2021	8.46	8.8028	8.8028	8.8028	0.3428	0.3428	0.3428	0.0405	0.0405	0.0405
2022	8.16	8.8336	8.8336	8.8336	0.6736	0.6736	0.6736	0.0825	0.0825	0.0825

TABLE 2.6 Recorded Error for Each of the Numerical Methods $\mu = 0.0040$

		Obtained Results			*Absolute Error = |Exact Value − Approximated Value|*			*Relative Error = Absolute Error/Exact Value*		
Year	**Exact Value (RM)**	**Runge-Kutta 4th-Order Method**	**3-Step P-C Method**	**Adams-Bashforth Method**	**Runge-Kutta 4th-Order Method**	**3-Step P-C Method**	**Adams-Bashforth Method**	**Runge-Kutta 4th-Order Method**	**3-Step P-C Method**	**Adams-Bashforth Method**
2011	8.50	8.5	8.5	8.5	0	0	0	0	0	0
2012	8.58	8.5341	8.5341	8.5341	0.0459	0.0459	0.0459	0.0053	0.0053	0.0053
2013	9.20	8.5683	8.5683	8.5683	0.6317	0.6317	0.6317	0.0686	0.0686	0.0686
2014	9.10	8.6026	8.6026	8.6026	0.4974	0.4974	0.4974	0.0546	0.0546	0.0546
2015	8.72	8.6371	8.6371	8.6371	0.0829	0.0829	0.0829	0.0095	0.0095	0.0095
2016	8.40	8.6717	8.6717	8.6717	0.2717	0.2717	0.2717	0.0323	0.0323	0.0323
2017	8.20	8.7065	8.7065	8.7065	0.5065	0.5065	0.5065	0.0617	0.0617	0.0617
2018	9.50	8.7414	8.7414	8.7414	1.0586	1.0586	1.0586	0.1080	0.1080	0.1080
2019	9.50	8.7764	8.7764	8.7764	0.7236	0.7236	0.7236	0.0761	0.0761	0.0761
2020	8.36	8.8116	8.8116	8.8116	0.4051	0.7648	0.7648	0.0540	0.0540	0.0540
2021	8.46	8.8469	8.8469	8.8469	0.3351	0.4051	0.4051	0.0457	0.0457	0.0457
2022	8.16	8.8824	8.8824	8.8824	0.6651	0.3351	0.3351	0.0885	0.0885	0.0885

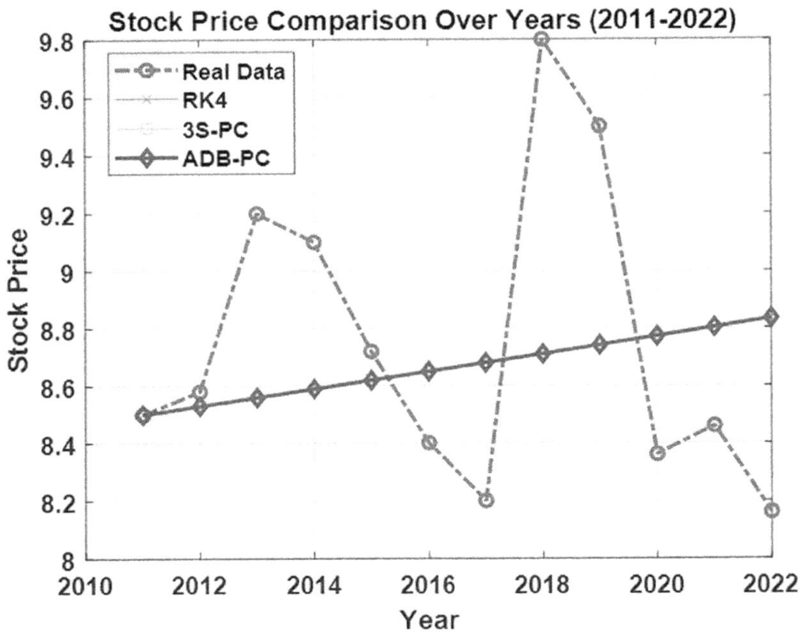

FIGURE 2.2 Graphical results of the three methods and exact stock price at $\mu = 0.0035$.

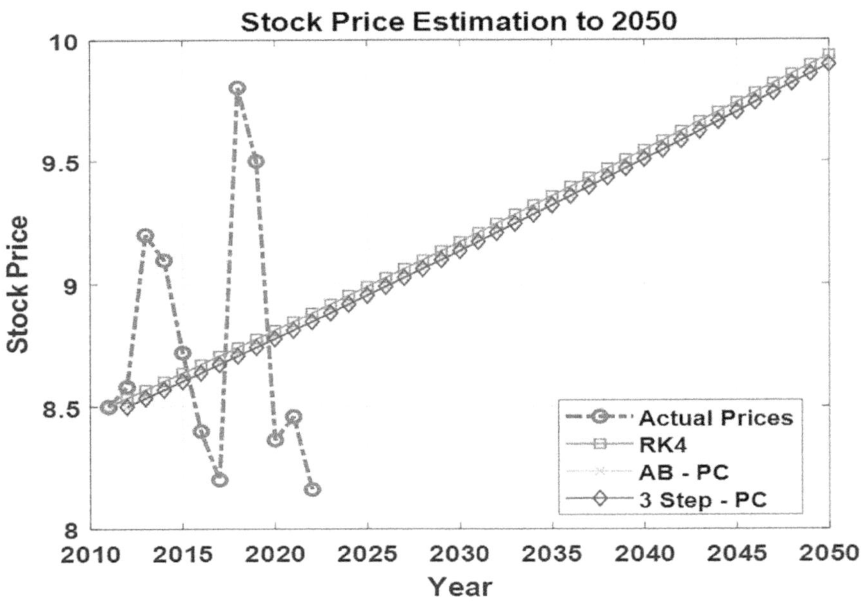

FIGURE 2.3 Graphical results of the three methods and exact stock price at $\mu = 0.0040$.

display the stock prices on the y-axis against time on the x-axis, highlighting the differences and similarities in the trajectory of predicted prices.

2.8 DISCUSSION OF NUMERICAL RESULTS

Figure 2.2 shows graphical representations of the solutions derived through the utilisation of all three numerical methods in comparison to the exact stock price from 2011 to 2022, with the value of the μ taken to be 0.0035. Simultaneously, Figure 2.3 portrays a similar depiction of this information, with $\mu = 0.0040$. Notably, the anticipated stock prices produced by all three methods exhibit a remarkable degree of similarity, demonstrating minimal discernible differences among them. It is evident from the data in Table 2.3 that the prediction made using $\mu = 0.0035$ produces smaller stock price values in contrast to the predicted stock price obtained, $\mu = 0.0040$, as demonstrated in Table 2.3. This can be attributed to the fact that $\mu = 0.0035$ produces smaller errors compared to the predictions made with $\mu = 0.0040$; thus we can infer that the choice of $\mu = 0.0035$ demonstrated a better choice drift value, proceeding further to prognosticate stock prices for Maybank spanning from 2011 to 2050 employing $\mu = 0.0035$ exhibit a lower forecast and reduced error compared to the error gotten using $\mu = 0.0040$. The three numerical methods follow a consistent trend and yield closely related predicted stock values. Furthermore, from Figure 2.3 it becomes apparent that the choice of $\mu = 0.0035$ outperforms the choice of $\mu = 0.0040$ in effectively capturing the underlying information present in the empirical data of Maybank. Notably, the convergence of the three numerical techniques signifies a sustained adherence to the long-term trends characterising Maybank's stock price fluctuations.

In Figure 2.4 the graphical representations of predicted values until 2050 using all three numerical methods using $\mu = 0.0035$ are displayed, while Figure 2.5 visualises the predicted stock price when $\mu = 0.0040$ is applied. These figures exhibit notable similarities to the previously discussed figures, with the only distinction being an exploration of the initial trend-capturing quality exhibited by each numerical method. A keen observation of these figures reveals that both approaches manifest analogous trends extending until 2050. This phenomenon underscores the pivotal role of using an accurate drift term value in guaranteeing the quality of predictions using GBM.

From the results obtained in our study, we can deduce that irrespective of the differing mathematical approaches employed, all three

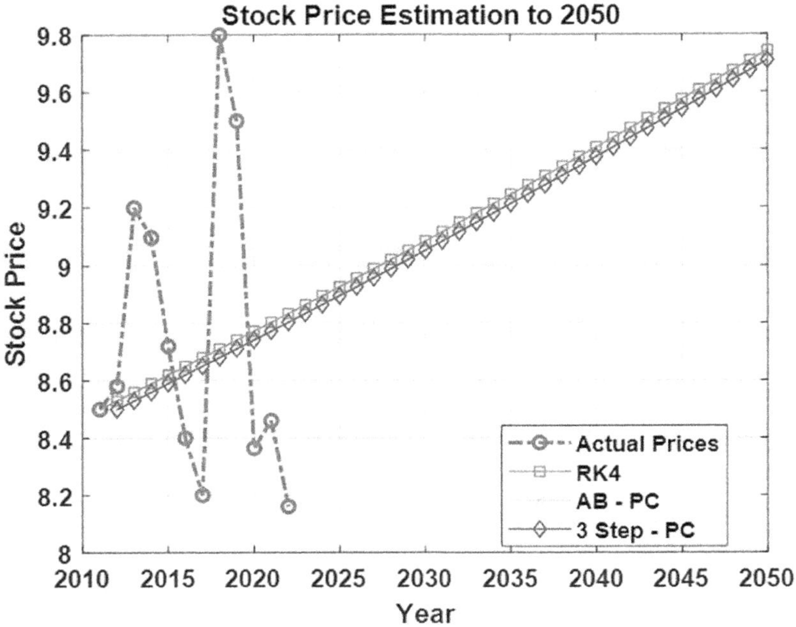

FIGURE 2.4 Graphical representation of predicted stock prices until 2050 at $\mu = 0.0035$.

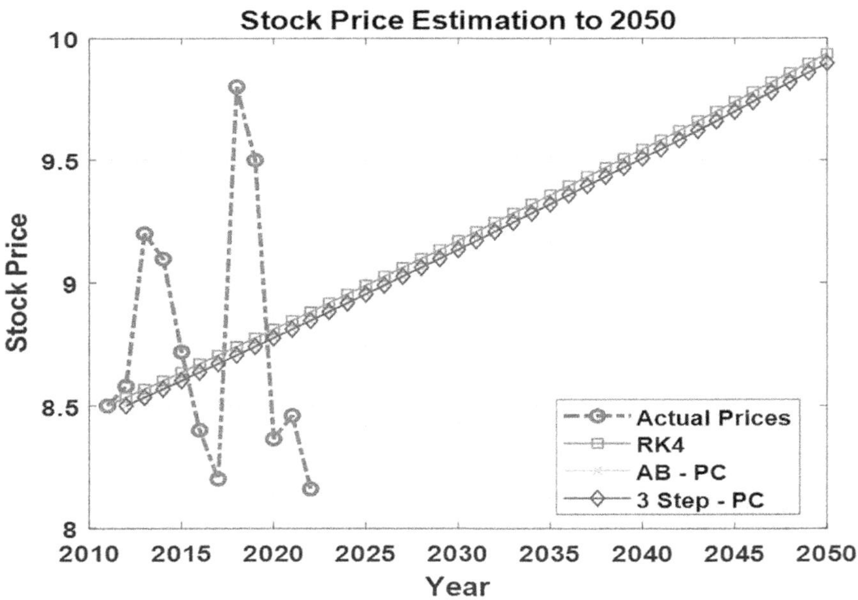

FIGURE 2.5 Graphical representation of predicted stock prices until 2050 at $\mu = 0.0040$.

methods concur on the general trajectory of Maybank's predicted stock prices over the extended term. This consistency enhances the reliability and credibility of the predictions, indicating a high degree of dependability. The convergence of all three methods towards a common value for the year 2050 underscores the stability inherent in the predictions. This stability implies that the models are relatively impervious to minor fluctuations or deviations in the input data or the chosen numerical technique, and the incorporation of a specific MGR exerts a profound influence on the overall upward or downward movement of the projected stock prices. The synergy between this growth rate and historical data serves as the driving force behind these predictions. It is imperative to recognise that variations in this growth rate can yield disparate predicted outcomes, highlighting the pivotal role of selecting the appropriate growth rate in shaping the long-term trajectory of Maybank's stock prices.

2.9 CONCLUSION

This chapter highlights the role of prudence and wise decision-making when contemplating stock price projections, providing insightful information for potential investors. All three numerical methods consistently produce nearly identical results when the volatility is set to zero. This result is not surprising because volatility, which represents the randomness and fluctuations in stock prices, is eliminated from the equation. Consequently, the predictions remain stable and closely aligned across the methods. However, it is essential for investors to use caution when choosing the appropriate method for numerical simulation. The forecasts made by the models, which are based on historical data and mathematical algorithms, do not take into consideration unforeseen factors that might influence stock market behaviour in the actual world, such as market policy, geopolitical events and economic situations. Unanticipated occurrences like pandemics can also cause large variations that go beyond the capabilities of the models. The convergence of forecasts is encouraging, but it has to be supported by a thorough risk analysis and examination of broader market circumstances. With an MGR from the datasets, the convergence of forecasted stock prices using the three techniques offers a consistent and steady forecast for Maybank's stock values up to the year 2050. Although these forecasts are a useful tool, they may be included in a more thorough investment study.

REFERENCES

Al-Harthy, M. H. (2007). Stochastic oil price models: Comparison and impact. *The Engineering Economist, 52*(3), 269–284. https://doi.org/10.1080/00137910701503944

Asif Arefin, M. (2020). A comparative exploration on different numerical methods for solving ordinary differential equations. *Journal of Mechanics of Continua and Mathematical Sciences, 15*(12). https://doi.org/10.26782/jmcms.2020.12.00001

Black, F., & Scholes, M. (1973). The pricing of options and corporate liabilities. *Journal of Political Economy, 81*(3), 637–654. https://doi.org/10.1086/260062

Brătian, V., Acu, A.-M., Mihaiu, D. M., & Şerban, R.-A. (2022). Geometric Brownian Motion (GBM) of stock indexes and financial market uncertainty in the context of non-crisis and financial crisis scenarios. *Mathematics, 10*(3), 309. https://doi.org/10.3390/math10030309

Campbell, J. Y., Lo, A. W., & MacKinlay, A. C. (1997). *The econometrics of financial markets.* Princeton University Press. https://doi.org/10.1515/9781400830213

Dash, M. (2019). Testing the random walk hypothesis in the Indian stock market using ARIMA modelling. *Journal of Applied Management and Investments, 8*(2), 71–77.

Delgado, J. E. R. (2019). Île de la passion in the transpacific framework. *Advances in Historical Studies, 8*(2), 91–98.

Djeutcha, E., Njamen Njomen, D. A., & Fono, L.-A. (2019). Solving arbitrage problem on the financial market under the mixed fractional Brownian motion with hurst parameter H ∈]1/2,3/4[. *Journal of Mathematics Research, 11*(1), 76. https://doi.org/10.5539/jmr.v11n1p76

Fama, E. F. (1970). Efficient capital markets: A review of theory and empirical work. *The Journal of Finance, 25*(2), 383. https://doi.org/10.2307/2325486

Farida Agustini, W., Affianti, I. R., & Putri, E. R. (2018). Stock price prediction using geometric Brownian motion. *Journal of Physics: Conference Series, 974*, 012047. https://doi.org/10.1088/1742-6596/974/1/012047

Islam, Md. A. (2015). A comparative study on numerical solutions of Initial Value Problems (IVP) for Ordinary Differential Equations (ODE) with Euler and Runge Kutta methods. *American Journal of Computational Mathematics, 5*(03), 393–404. https://doi.org/10.4236/ajcm.2015.53034

Jabbarzadeh, A., Shavvalpour, S., Khanjarpanah, H., & Dourvash, D. (2016). A multiple-criteria approach for forecasting stock price direction: Nonlinear probability models with application in S&P 500 index. *International Journal of Applied Engineering Research, 11*(6), 3870–3878.

Konechnaya, O., Schwanen, C., & Schwarzbauer, J. (2021). Application of multistep approach for comprehensive identification of microplastic particles in diverse sediment samples. *Water Science and Technology, 83*(3), 532–542. https://doi.org/10.2166/wst.2020.600

Madziwa, L., Pillalamarry, M., & Chatterjee, S. (2022). Gold price forecasting using multivariate stochastic model. *Resources Policy, 76*, 102544. https://doi.org/10.1016/j.resourpol.2021.102544

Mahdiani, M. R., & Khamehchi, E. (2016). A modified neural network model for predicting the crude oil price. *Intellectual Economics, 10*(2), 71–77. https://doi.org/10.1016/j.intele.2017.02.001

Mensah, E. T., Boateng, A., Frempong, N. K., & Maposa, D. (2023). Simulating stock prices using geometric Brownian motion model under normal and convoluted distributional assumptions. *Scientific African, 19*, e01556. https://doi.org/10.1016/j.sciaf.2023.e01556

Nemes, M. D., & Butoi, A. (2013). Data mining on Romanian stock market using neural networks for price prediction. *Informatica Economica, 17*(3/2013), 125–136. https://doi.org/10.12948/issn14531305/17.3.2013.11

Shahi, T. B., Shrestha, A., Neupane, A., & Guo, W. (2020). Stock price forecasting with deep learning: A comparative study. *Mathematics, 8*(9), 1441. https://doi.org/10.3390/math8091441

Zahedi, J., & Rounaghi, M. M. (2015). Application of artificial neural network models and principal component analysis method in predicting stock prices on Tehran stock exchange. *Physica A: Statistical Mechanics and Its Applications, 438*, 178–187. https://doi.org/10.1016/j.physa.2015.06.033

CHAPTER 3

Collective Dynamics in a Random Network of Pulse-Coupled Oscillators with Heterogeneous Intrinsic Frequencies

Afifurrahman

3.1 INTRODUCTION

Network oscillator theory is pivotal in understanding synchronisation phenomena across various biological and artificial systems. It focuses on how interconnected oscillatory units, such as neurons or electronic circuits, achieve coordinated rhythmic activity. By modelling interactions through coupling mechanisms, such as synaptic connections or physical coupling in circuits, the theory explains emergent synchronisation patterns (Ashwin et al., 2016). This framework elucidates how network topology, coupling strength and individual oscillator properties collectively influence synchronisation. Insights gleaned from network oscillator theory are essential for comprehending brain dynamics, collective behaviours

DOI: 10.1201/9781032678573-3

in social systems (Proskurnikov & Tempo, 2018) and designing robust communication networks (Chen et al., 2005), offering fundamental principles for both theoretical exploration and practical applications in diverse fields.

The early investigations into collective phenomena, e.g., synchronisation, dated back when physicists studying coupled oscillators, such as pendulums and mechanical systems. The work of Christiaan Huygens in the 17th century, who observed synchronisation of pendulum clocks, laid foundational concepts (Peña Ramirez et al., 2016). Later, the mathematical models like the Kuramoto model, introduced by Yoshiki Kuramoto in 1975, provided a formal framework to study synchronisation in large populations of oscillators (Kuramoto, 1984). These models use differential equations to describe how phase differences evolve over time due to coupling mechanisms.

Collective dynamics in networks of pulse-coupled oscillators with diverse intrinsic frequencies have been extensively studied to understand how synchronisation emerges in complex systems. In these networks, each oscillator possesses a unique intrinsic frequency (Acebrón et al., 2005), and interactions occur through brief pulses exchanged between neighbouring oscillators (Canavier, 2013). This setup mimics various biological and physical systems, from neuronal firing in the brain to synchronisation in firefly populations.

The past research has demonstrated that even in networks where oscillators have significantly different intrinsic frequencies, they can synchronise under certain conditions (Mirollo & Strogatz, 1990). This phenomenon, known as phase locking, occurs when oscillators adjust their phases relative to each other despite their frequency disparities. The coupling strength with mutual excitation between oscillators plays a crucial role, influencing how quickly synchronisation occurs and the stability of synchronised states.

The variations in neuron properties, such as ion channel distributions and membrane potentials, might influence collective oscillatory behaviour in neural networks that inhibit each other (White et al., 1998). Through computational models and experiments, the mutual inhibition can synchronise neuronal activity despite heterogeneity. The study highlights how inhibitory interactions shape network dynamics, offering insights into how diverse neural populations coordinate their activity, which is critical for understanding brain function and potential implications for neurological disorders characterised by dysregulated synchronisation.

Further investigations in Pikovsky et al. (2001) explore the effects of network topology on synchronisation. They found that the structure of the network—whether regular, random or small world—can significantly impact synchronisation dynamics. Small-world networks, for example, promote faster synchronisation compared to regular networks due to their balance of local clustering and long-range connections.

Moreover, the presence of diverse intrinsic frequencies introduces richer dynamics as studied in a developed framework using the Kuramoto model (Ott & Antonsen, 2008). The work showed that networks of oscillators with heterogeneous frequencies can exhibit complex behaviours such as frequency entrainment and cluster synchronisation. These findings underscore the robustness and adaptability of synchronisation in diverse networks, highlighting the intricate interplay between intrinsic properties and network connectivity.

In this chapter, we expand the investigation on the effects of heterogeneous intrinsic frequencies in network of excitatory and inhibitory neurons interacting through finite-width pulses which have been inspired by previous studies (Afifurrahman et al., 2020, 2021). Neurons in the brain exhibit a range of intrinsic frequencies due to variations in ion channel kinetics, membrane properties and other biophysical factors (Ito, 2008). By assuming a uniform distribution of frequencies across a population of neurons, computational models can simulate this diversity more accurately (Tao et al., 2018; Ullner & Politi, 2016). This approach helps capture the statistical properties of neuronal networks, influencing how synchronisation, information processing and other emergent behaviours unfold in response to external stimuli and network dynamics.

Due to the complexity of the model, we limit ourselves on numerical simulations. However, the computation is accompanied by a finite-size scaling analysis to justify the behaviour of larger network sizes. The first result is that the irregular collective behaviour is observed when the balance state is maintained with the incorporation of internal frequency dispersions. Such a regime persists as the network size is increased. For a relatively small coupling strength, the perfect synchronous dynamics vanished when the excitatory pulse-width is shorter than the inhibitory one, which contrast with the research findings in Afifurrahman et al. (2020, 2021). This result confirms the critical role of neurons internal frequency in shaping neural network dynamics.

The chapter is organised as follows. In section 3.2 we define the mathematical model including the phase-response curve used for numerical simulations. In section 3.3, we introduce the methods and indicators to characterise the collective dynamics. In sections 3.4 and 3.5, we provide the evidence of irregular collective dynamics and synchrony degeneration in the context of heterogeneous neural networks, respectively. Section 3.6 is devoted to discussion, and section 3.7 is conclusion.

3.2 MODEL

The object of study is N phase oscillators, also being referred to as neurons, composed of $N_E = bN$ excitatory and $N_I = (1-b)N$ inhibitory neurons. The synaptic connections are represented by random matrix $\mathbf{G} = (G_{j,k})$: if $k \to j$ is active then $G_{j,k} = 1$, otherwise $G_{j,k} = 0$ where $1 \le j,k \le N$. The network structure is without autapse, i.e., $G_{j,j} = 0$.

All neurons receive the same number ℓ_E and ℓ_I of excitatory and inhibitory connections (obviously, altogether $\ell = \ell_E + \ell_I$). Further, we assume massive connectivity in which ℓ is proportional to N, i.e., $\ell = cN$ where c is the connection probability. Figure 3.1a displays the network model studied in this chapter.

Each neuron is characterised by phase variable $\phi_j \le 1$. We generalise the homogeneous model for both excitatory and inhibitory neurons introduced in Afifurrahman et al. (2021) to a heterogeneous case by adjusting the intrinsic frequency ω_j for each individual neuron. Mathematically,

$$\begin{cases} \dot{\phi}_j = \omega_j + \dfrac{\varepsilon}{\sqrt{K}} \Gamma(\phi_j)(E_j - I_j) \\ \dot{E}_j = -\alpha\left(E_j - \sum_n G_{j,k} P_k \delta\left(t - t_{n(k)}\right)\right) \\ \dot{I}_j = -\beta\left(I_j - g\sum_n G_{j,k}(1 - P_k)\delta\left(t - t_{n(k)}\right)\right) \end{cases} \tag{3.1}$$

where ε is the coupling strength, $\Gamma(\phi)$ represents the phase-response curve and E_j and I_j are the excitatory and inhibitory fields, respectively, perceived by the neuron.

The parameters α and β denote the inverse pulse-width of excitatory and inhibitory exponential pulses, while g accounts for a relative strength of inhibition compared to the excitation and $\delta(t)$ is the delta function. Finally, the vector P_k separates excitatory from inhibitory ones: if the kth neuron is excitatory, then $P_k = 1$, otherwise $P_k = 0$.

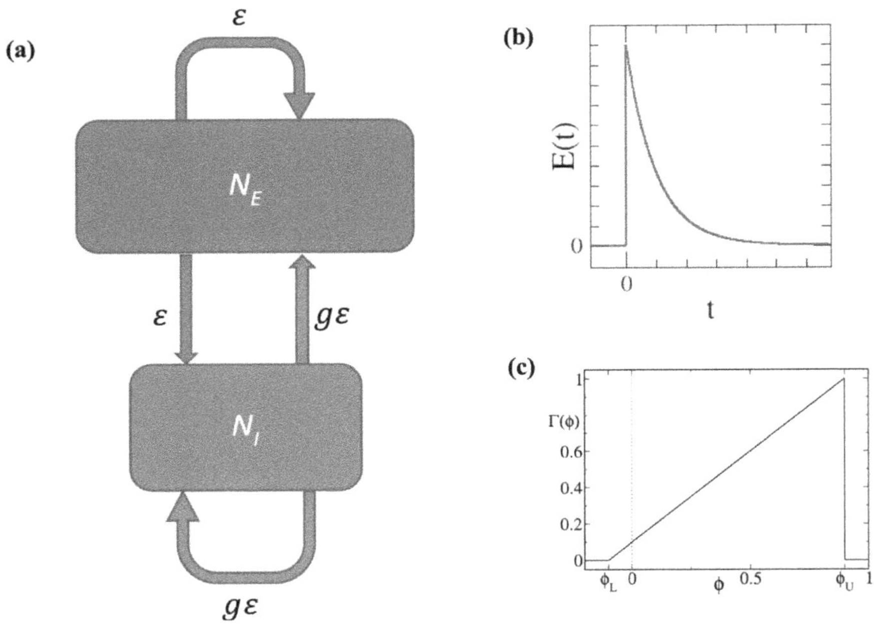

FIGURE 3.1 (a) Network model for two populations of non-identical excitatory and inhibitory neurons, (b) shape of exponential pulse and (c) shape of simplified phase-response curve. The dotted line denotes the reset value $\phi_\circ = 0$.

Source: Afifurrahman et al. (2020, 2021).

The dynamics of differential Eq. (3.1) can be summarised as follows: whenever ϕ_k passes a threshold $\phi_* = 1$, it is reset to $\phi_\circ = 0$ and enters a refractory time t_r during which it stands and is insensitive to the action of $E_j(I_j)$. Simultaneously, the fields $E_j(I_j)$ are activated as an effect of nth pulse sent by neuron k at time $t_{n(k)}$. The shape of a single pulse is displayed in Figure 3.1b.

The following piecewise linear function defines the phase-response curve to mimic the exponential shape in the leaky integrate-and-fire neurons with delay:

$$\Gamma\left(\phi_j\right) = \begin{cases} \phi_j - \phi_L, & \text{if } \phi_L < \phi_j < \phi_U \\ 0, & \text{otherwise,} \end{cases} \tag{3.2}$$

where ϕ_L and ϕ_U a the lower and upper bounds of the phase-response curve, respectively. The resulting curve is given in Figure 3.1c. For a numerical simulation, we fix $\phi_L = -0.1$ and $\phi_U = 0.9$.

3.3 NUMERICAL ANALYSIS AND STATISTICS

The network dynamics is simulated by implementing the Euler algorithm with the integration time-step $\Delta t = 10^{-3}$. The length of integration time is $t = 10000$ time unit; 4000 of the simulation time has been discarded to let enough of the system converge to the attractor. The initial conditions for the phases $\phi_j(0)$ are chosen randomly from a uniform distribution in the unit interval $[0,1]$, while the fields $E_j(0)$ and $I_j(0)$ are all initially set to zero. The frequency ω_j is drawn from a uniform distribution of $D(\omega)$ centred around $\omega_0 = 1$ with a small dispersion $\Delta_\omega \ll 1$. The focal parameter in this work is Δ_ω. Notice that in the limit of $\Delta_\omega \to 0$, all neurons share the same frequency, i.e., $\omega_j \to 1$ for all j. We expect the emerging dynamical regimes are consistently the same as presented in Afifurrahman et al. (2021) for $\Delta_\omega = 0$.

In all cases we work with $b = 0.8$ and $c = 0.1$ as often used in the existing literatures (Ostojic, 2014). We rescale the coupling term by $1/\sqrt{K}$ to maintain the balanced regime in the limit $K, N \to \infty$. A small degree of unbalance between excitation and inhibition has been introduced by setting $g = 4 + \sqrt{1000/K}$ (Ullner et al., 2018). We also assume a weak coupling strength ε in the order of 10^{-1} to ensure the validity of phase reduction principle. The values of $\alpha = 100, \beta \in [20,110]$ and $t_r = 0.03$ time units have been selected according to the computational experiments (Afifurrahman et al., 2021). In order to test the finite-size effects of the emerging dynamics, the network size N is varied.

The spiking activity of population of neurons is quantified by the *mean firing rate*

$$\nu = \lim_{t \to \infty} \frac{1}{tN} \sum_{j=1}^{N} n_j(t) \tag{3.3}$$

where $n_j(t)$ denotes the number of spikes emitted by the neuron j over time t. The inter-spike interval (ISI) of a single neuron is time difference of two consecutive spikes, i.e., $t_{n+1} - t_n$. *Mean coefficient of variation* $\overline{cv}$ is a statistical measure based on the variability of ISI. It is estimated by a ratio

$$\overline{cv} = \frac{1}{N} \sum_{j=1}^{N} \frac{\sigma_j}{\tau_j} \tag{3.4}$$

where σ_j is a standard deviation of ISI and τ_j is mean ISI of jth neuron. If $\overline{cv} < 1$, then the spike train $\{t_n\}$ (and hence ISI) is relatively regular, while if $\overline{cv} > 1$ then the neurons show bursting activity.

To quantify the collective dynamics in large neuronal networks, it is necessary to introduce *order parameter* R. It is defined as

$$R^2 \equiv \frac{\overline{\langle \phi \rangle^2} - \overline{\langle \phi \rangle}^2}{\langle \overline{\phi^2} - \overline{\phi}^2 \rangle} \tag{3.5}$$

where the angular bracket $\langle \circ \rangle$ represents an ensemble average, while the over-bar is a time average. If all neurons behave exactly in the same way (perfect synchronisation), then $R = 1$. If instead, they are independent (asynchronous state), then $R \approx 1/\sqrt{N}$.

The regime where $0 < R < 1$ is known as partial synchronisation: in this case, a fraction of neurons share common dynamics. An instance of partial synchrony is the so-called collective irregular dynamics (CID). The peculiarity of CID is that the ensemble average of the global variable, e.g., average membrane potential (or in this context, the average phase $\langle \phi \rangle$), fluctuates irregularly in time. Several mathematical properties accompanying the emergence of CID in networks of oscillators: (1) $\langle \phi \rangle$ shows a stochastic-like dynamics, (2) the power spectra of $\langle \phi \rangle$ have a broad shape and are independent of the system size N and (3) the instantaneous distribution of the phase changes with time, in contrast to that of standard asynchronous regime.

3.4 COLLECTIVE IRREGULAR DYNAMICS (CID)

In this section, we consider CID in the presence of heterogeneous intrinsic frequency. Such regime is characterised by an irregular fluctuation of the global variable with respect to time. Our first evidence is displayed in Figure 3.2a where we have computed the time series of average phase $\langle \phi \rangle (t) = \frac{1}{N} \sum_{j=0}^{N} \phi_j (t)$ for the same pulse-widths $\alpha = \beta$ and nonzero frequency dispersion $(\Delta_\omega \neq 0)$.

As seen in Figure 3.2a, $\langle \phi \rangle (t)$ shows stochastic-like oscillation, which confirms the irregular pattern of collective behaviour. The raster plot of 100 oscillators in Figure 3.2b suggests less coordinated spiking times, a supportive indication of a weakly synchronised state. More quantitative representation is given by the instantaneous probability distribution $\rho(\phi)$ in Figure 3.2c. $\rho(\phi)$ is plotted at time: $t = 2000$ (red) and $t = 2800$ (green). The peak at zero is because some oscillators still stand at the refractory period. At variance with the standard asynchronous regime, the shape of

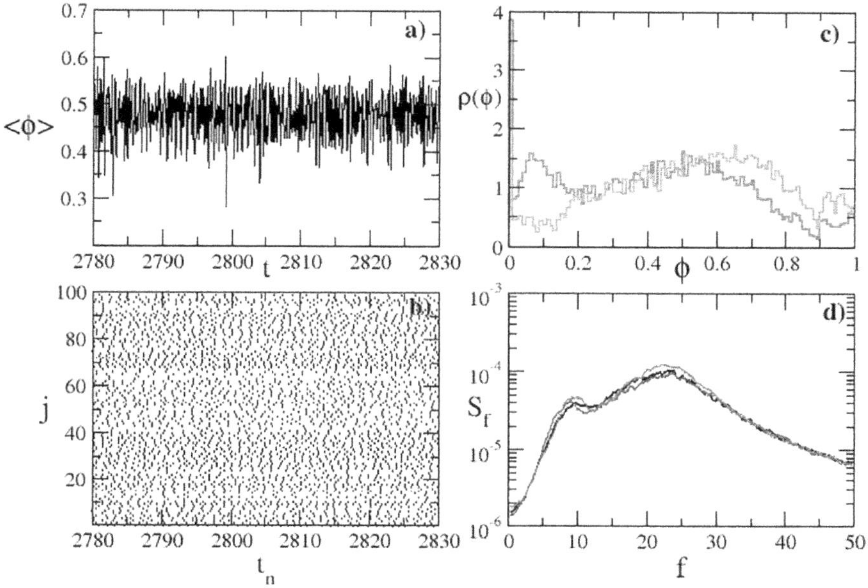

FIGURE 3.2 CID properties for $\Delta_\omega = 0.6$, $\varepsilon = 0.3$, $\alpha = 100$ and $\beta = 100$. (a) Time series of mean-field ϕ computed for $N = 10000$. (b) Raster plot of spiking times t_n 100 oscillators out of $N = 10000$. (c) Instantaneous probability distribution of the phases $\rho(\phi)$ at two different time points: $t = 2000$ (red) and $t = 2800$ (green). (d) Power spectra of the average phase S_f as a function of frequency f for different network sizes: $N = 10000$ (black), $N = 20000$ (blue) and $N = 40000$ (magenta).

$\rho(\phi)$ changes with time. Finally, in Figure 3.2d we report the outcome of simulations to test the finite-size effects. The shape of power spectrum of average phase is sufficiently wide under the curve $S_f(\langle\phi\rangle)$ that reflects a complex nature of CID. Meanwhile, $S_f(\langle\phi\rangle)$ is insensitive to the network sizes N: 10000 (black), 20000 (blue) and 40000 (magenta), which confirms the persistence of CID in the heterogeneous neuron model (3.1).

3.5 SYNCHRONY DEGRADATION

As a consequence of heterogeneous frequency, the dynamical state of network of oscillators may switch from perfect to partial synchronisation when the excitatory pulse-width is shorter than inhibitory one. We demonstrate such mechanism upon computing the indicators for $\varepsilon = 0.3$ and $\Delta_\omega \in [0, 0.6]$. At first, we perform an analysis of the neuron's firing activity

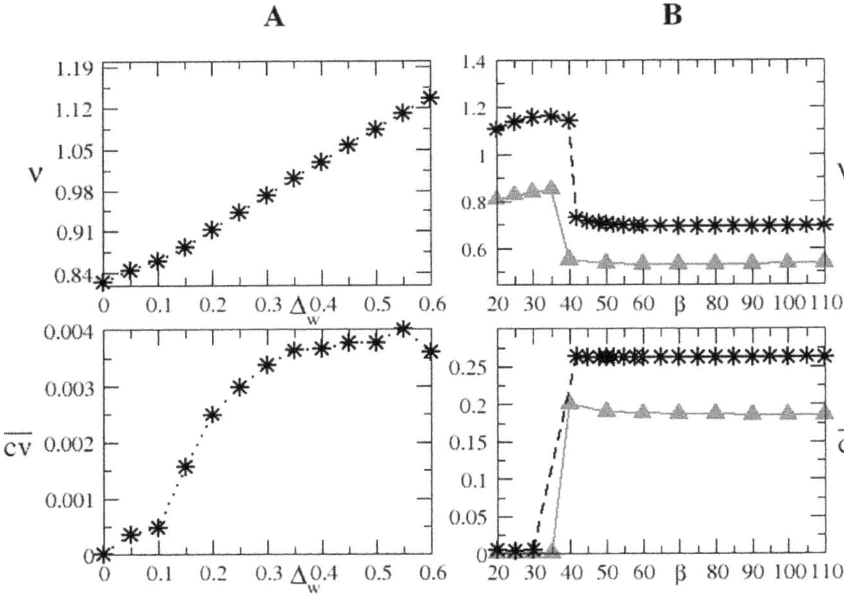

FIGURE 3.3 (a) Mean firing rate ν and mean coefficient of variation $\overline{cv}$ versus the width of frequency distribution Δ_ω when $\beta = 25$. (b) Mean firing rate ν and mean coefficient of variation $\overline{cv}$ versus β: magenta triangles correspond to homogeneous neuronal network ($\Delta_\omega = 0$) while black stars correspond to heterogeneous neurons ($\Delta_\omega = 0.6$). All simulations are performed for $N = 10000$, $\varepsilon = 0.3, \alpha = 100$.

by calculating the mean firing rate, ν, and mean coefficient of variation, $\overline{cv}$, for $N = 10000$. The numerical results are displayed in Figure 3.3.

More specifically, Figure 3.3a displays ν and $\overline{cv}$ as a function of Δ_ω when $\alpha = 100$, $\beta = 25$. The estimate values for ν and $\overline{cv}$ at lower bound $\Delta_\omega = 0$ match the corresponding indicators for the case of homogeneous neurons (see Figure 3.5a in Afifurrahman et al., 2021). Upon increasing Δ_ω, both quantities show a moderate growth, meaning that the width of frequency dispersion affects the firing activities of the whole network, though the regularity of the neuron's ISI is preserved as depicted by small magnitude of $\overline{cv}$ in the order of 10^{-3}.

A more comprehensive description is shown in Figure 3.3b, where the measures ν and $\overline{cv}$ are plotted against β for the two network setups, i.e., $\Delta_\omega = 0$ and $\Delta_\omega = 0.6$ corresponding to the homogeneous and heterogeneous neuronal network, respectively. The indicators reveal a significant impact of the heterogeneity to network dynamics for wide range of β. There, one can see the mean firing rate for the heterogeneous network is larger than that of the homogeneous one when $\beta \in [20,110]$. A similar scenario has also been detected for the mean coefficient of variations except for $\beta \in [20,40)$: the $\overline{cv}$ for homogeneous and heterogeneous system is near zero and are almost indistinguishable. In such a case, the regular ISI of an individual oscillator is preserved.

Figure 3.4 display the spiking times t_n of neurons (a,c) and the distribution of the phase $\phi_j(t)$ (b,d). We have selected $\beta = 35$ while keeping another parameter set the same as in Figure 3.3. In the absence of

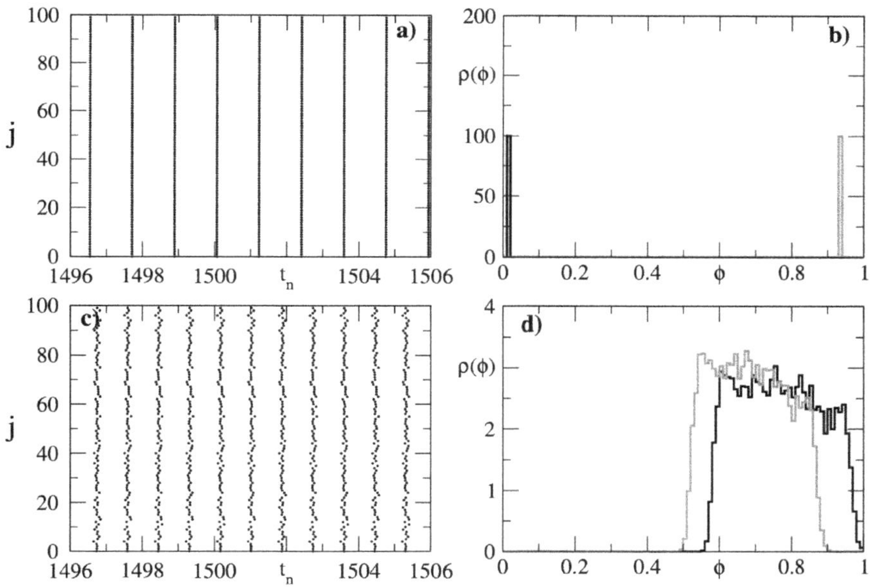

FIGURE 3.4 (a) Raster plot of the spiking times t_n for 100 oscillators out of $N = 10000$ shows the emergence of fully synchronised state when $\Delta_\omega = 0$. (b) Instantaneous probability distribution of the phases, $\rho(\phi)$, at two sampled time points: $t = 500$ (black) and $t = 1500$ (red) in the fully synchronous regime. (c) Raster plot of the spiking times t_n for 100 oscillators out of $N = 10000$ shows a synchrony breaking phenomenon when the finite-width of frequency $\Delta_\omega = 0.6$ has been imposed. (d) Instantaneous probability distribution of the phases $\rho(\phi)$ at two sampled time points: $t = 500$ (black) and $t = 1500$ (red) in the highly synchronised state. All simulations are performed for $\alpha = 100$, $\beta = 35$ and $N = 10000$ neurons. The probability distribution is normalised in such a way that the area under the curve is 1.

heterogeneous intrinsic frequency ($\Delta_\omega = 0$), the first 100 neurons out of $N = 10000$ are firing exactly at the same time (Figure 3.4a). The distribution, $\rho(\phi)$, confirms that 10000 neurons are perfectly synchronised since the phases are all localised at the same position for each time point, i.e., $\phi_j(t) = \phi(t)$ for all j and $t \to \infty$ (see Figure 3.4b as an example). Unfortunately, perfect synchronous dynamics is annihilated when the heterogeneity is present. As for instance, while choosing $\Delta_\omega = 0.6$ the spiking times of 100 neurons become highly, but not perfectly, synchronised (Figure 3.4c). This finding is further clarified by the distribution $\rho(\phi)$: the phase $\phi_j(t)$ for each sampling time t are distributed over a finite interval $\phi(t) \in [0.5, 1]$ in Figure 3.4d.

Further analysis of synchronous firing behaviour is carried out by computing the synchrony measure R with respect to the parameter $\Delta_\omega \in [0, 0.6]$. We fix $\beta = 35$ as a reference point in which the perfect synchronous state is stable, i.e., $R(\Delta_\omega = 0) = 1$ Afifurrahman et al. (2020, 2021). While Δ_ω is increased, the quantity R declines progressively below 1, meaning that the whole system behaves in near-perfect synchronisation (Figure 3.5a). Such a regime persists for fixed $\Delta_\omega = 0.6$ upon shortening the pulse-width in the range of $\beta \in [20, 110]$: see the black stars in Figure 3.5b. In comparison to the heterogeneous networks, we also plot the same indicator $R(\beta)$ for $\Delta_\omega = 0$. The particularity of homogeneous network is that the neurons are perfectly synchronised in the region of $\beta \in [20, 40)$, accompanied by a sudden jump around $\beta_c \approx 40$, which a

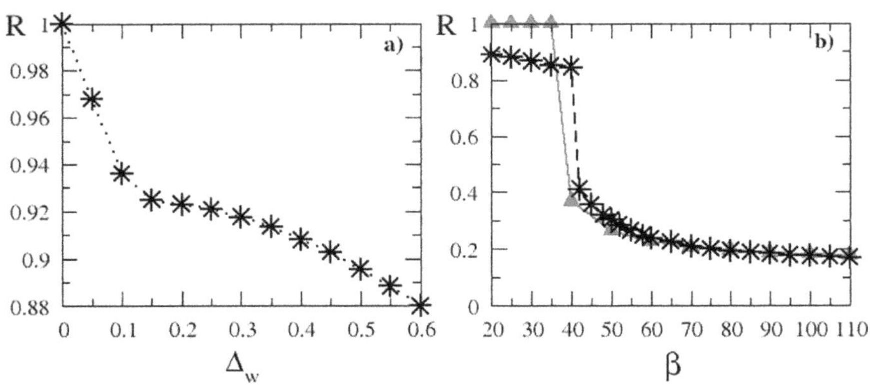

FIGURE 3.5 (a) Order parameter R as a function of Δ_ω. (b) Order parameter R as a function of β. All parameter sets are the same as in Figure 3.3.

seemingly like first-order transition. Above $\beta = 40$ the collective dynamics is quite similar to that of heterogeneous neuronal network. There, one can see that the synchrony measure R is relatively low, signalling a weak synchronisation amongst neurons.

A sudden jump around $\beta_c \approx 40$ on the mean firing rate and mean coefficient of variations (see Figure 3.3b) and order parameter (Figure 3.5b) is clear indication of first-order (discontinuous) phase transition. It occurs due to the relative width of excitatory and inhibitory pulses. In the case of a homogeneous neural network, the scenario is accompanied by the coexistence between CID and fully synchronous state (Afifurrahman et al., 2021). In the present heterogeneous model, we expect a similar scenario, except that the fully synchronous firing dynamics no longer exist.

3.6 DISCUSSION

In this chapter we have investigated numerically the emergence of collective dynamics in the balanced spiking neurons interacting through finite pulses. Instead of having identical intrinsic frequency as studied (Afifurrahman et al., 2020, 2021) the neuronal model considered here consists of non-identical oscillators characterised by uniform frequency distribution $D(\omega) \sim (1, \Delta_\omega)$. This way, the model of coupled oscillators is more generic and relevant to the biophysical constraints of a cell, not only because the nature of action potential shape but also due to the diversity of neuron internal characteristics (Vera et al., 2020; Wang et al., 2018).

It is firmly understood the collective irregular dynamics in balanced neural networks arise from the delicate interplay between excitation and inhibition, such as those found in simple binary networks and spiking neurons (Ullner et al., 2018; Vreeswijk & Sompolinsky, 1998). In these networks, the strength and distribution of synaptic connections are finely tuned to maintain a state where excitation and inhibition are balanced, preventing runaway excitation or quiescence. This balance allows for the emergence of irregular firing patterns among neurons, where spikes occur asynchronously due to random fluctuations and interactions.

The prevalence of irregular dynamics in the present model suggests the generality of such a regime in the central nervous system. Compared to the massive bombardment of excitatory and inhibitory feedback, the dispersion of intrinsic frequency (regardless how large Δ_ω is) is likely having no significant effect on the network dynamics when the balanced regime is maintained as $N \to \infty$ (see Figure 3.2 as an example and compare it with Figure 3.3 in Afifurrahman et al. (2021): the balanced regime is achieved while choosing the symmetrical width of finite excitatory and inhibitory pulses, i.e., $\alpha^{-1} = \beta^{-1}$. Despite the role played by the neurons' intrinsic properties in controlling the computational aspect of neural circuits (Gast et al., 2024), in this case the network connectivity often plays a more significant role related to the emerging collective phenomenon in the brain. This is because it typically involves the adjustment of phase relationships among oscillators rather than matching their frequencies exactly (Pikovsky et al., 2001).

The heterogeneous internal frequency affects the network dynamics when the excitatory pulse-width is shorter than the inhibitory one ($\alpha^{-1} > \beta^{-1}$), highlighted by the change of the statistical measures ν and $\overline{cv}$ as a function of Δ_ω (see again Figure 3.3a). While enlarging the main parameter Δ_ω, one would expect an immense number of spikes is produced by a single oscillator which results in the high average firing rate. A tiny change in the mean coefficient of variations indicates the regularity of inter-spike interval is preserved as further clarified by the raster plot in Figure 3.4b. In contrast to the previous study (Afifurrahman et al., 2021), the fully synchronised state is no longer perceived in the current setup with a relatively small coupling strength, and so the order parameter is less than $R = 1$ (Figure 3.5a). This is because the frequency differences adjust the phase-lag when the oscillators are coupled, like those that have been observed in a minimal network of two Kuramoto-like oscillators (Afifurrahman, 2022). The finding extends beyond the effects of heterogeneity in mutually coupled inhibitory neurons studied by White et al. (1998), in a way that the interplay of excitatory and inhibitory finite pulses can promote synchronous firing patterns in the heterogeneous neural networks.

Another effect of heterogeneous frequency is the emergence of subthreshold activity that accompanies the synchrony degradation (data not shown). In this regime, neurons integrate synaptic inputs through graded potentials, such as excitatory and inhibitory postsynaptic potentials (EPSPs and IPSPs), without producing pulses (Gerstner et al., 2014). This phase is crucial for information processing, as it allows neurons to compute and integrate inputs in a graded manner (Engel et al., 2008). Subthreshold dynamics are influenced by ion channel conductance, synaptic input strengths, intrinsic membrane properties and neuronal resonance, shaping how neurons respond to varying stimuli and contributing to the computational flexibility of neural networks (Richardson et al., 2003; Yu et al., 2023).

The emergence of first-order transition in this study highlights the importance of finite-width pulses in shaping the network dynamics (see Figures 3.3b and 3.5b). The phenomena refer to the state where synchronisation of oscillators or dynamical systems in a network change abruptly from a state of asynchronous behaviour to a synchronised state (or vice versa), when the network parameters exceed a critical value. The interplay between inhibitory pulse-widths and excitatory ones is mainly responsible, which in this case is varying β while α is fixed, besides other global coupling parameters such as the coupling strength and network topology seen in the Kuramoto oscillators (Arola-Fernández et al., 2022) and the noise parameter in the stochastic leaky integrate-and-fire neurons (Afifurrahman et al., 2023). In this work, the evidence of bistability that often accompanies the discontinuous transitions has not been explored further, yet we expect it occurs like what has been observed in the previous study (Afifurrahman et al., 2021) with the exemption of fully synchronous regime.

3.7 CONCLUSION

To summarise, we have proposed a network model for two interacting populations of excitatory and inhibitory neurons and provided the evidence of the effects of heterogeneous internal frequency on the network dynamics. The overall finding highlights the dependence of an emerging scenario on the degree of unbalance which is characterised by a relative strength between excitatory and inhibitory pulse-widths. When the balance regime is perfectly maintained, the irregular collective dynamics are ubiquitous. Understanding collective irregular dynamics in balanced networks is crucial for unravelling how the brain processes and integrates information across large populations of neurons (La Camera et al., 2008). It also provides insights into neurological disorders characterised by disruptions in network balance, offering potential avenues for therapeutic interventions targeting network dynamics (Uhlhaas & Singer, 2012). On the other hand, the unbalance between excitation and inhibition results in degradation of fully synchronised state towards partial synchrony when the internal frequency diversity is incorporated. However, this is true only when the coupling strength is small: for future reference, ones may go beyond the small coupling to extrapolate the emerging dynamics. Finally, it is necessary to perform a stability analysis for the synchronous states to validate the numerical findings in this study.

REFERENCES

Acebrón, J. A., Bonilla, L. L., Pérez Vicente, C. J., Ritort, F., & Spigler, R. (2005). The Kuramoto model: A simple paradigm for synchronization phenomena. *Reviews of Modern Physics, 77*(1), 137–185. https://doi.org/10.1103/RevModPhys.77.137

Afifurrahman, A., Mohd, M. H., & Abdullah, F. A. (2023). The interplay of common noise and finite pulses on biological neurons. *Communication in Biomathematical Sciences, 6*(2), 145–155. https://doi.org/10.5614/cbms.2023.6.2.5

Afifurrahman, E. U. (2022). *Collective dynamics in a network of pulse-coupled oscillators* [Thesis, University of Aberdeen].

Afifurrahman, E. U., & Politi, A. (2020). Stability of synchronous states in sparse neuronal networks. *Nonlinear Dynamics, 102*(2), 733–743. https://doi.org/10.1007/s11071-020-05880-4

Afifurrahman, E. U., & Politi, A. (2021). Collective dynamics in the presence of finite-width pulses. *Chaos, 31*(4). https://doi.org/10.1063/5.0046691

Arola-Fernández, L., Faci-Lázaro, S., Skardal, P. S., Boghiu, E.-C., Gómez-Gardeñes, J., & Arenas, A. (2022). Emergence of explosive synchronization bombs in networks of oscillators. *Communications Physics, 5*(1), 264. https://doi.org/10.1038/s42005-022-01039-2

Ashwin, P., Coombes, S., & Nicks, R. (2016). Mathematical frameworks for oscillatory network dynamics in neuroscience. *The Journal of Mathematical Neuroscience, 6*(1), 2. https://doi.org/10.1186/s13408-015-0033-6

Canavier, C. C. (2013). Pulse-coupled oscillators. In D. Jaeger & R. Jung (Eds.), *Encyclopedia of computational neuroscience* (pp. 1–11). Springer. https://doi.org/10.1007/978-1-4614-7320-6_269-1

Chen, G., Fan, Z., & Li, X. (2005). Modelling the complex internet topology. In L. Kocarev & G. Vattay (Eds.), *Complex dynamics in communication networks* (pp. 213–234). Springer. https://doi.org/10.1007/10973509_9

Engel, T. A., Schimansky-Geier, L., Herz, A. V. M., Schreiber, S., & Erchova, I. (2008). Subthreshold membrane-potential resonances shape spike-train patterns in the entorhinal cortex. *Journal of Neurophysiology, 100*(3), 1576–1589. https://doi.org/10.1152/jn.01282.2007

Gast, R., Solla, S. A., & Kennedy, A. (2024). Neural heterogeneity controls computations in spiking neural networks. *Proceedings of the National Academy of Sciences, 121*(3), e2311885121. https://doi.org/10.1073/pnas.2311885121

Gerstner, W., Kistler, W. M., Naud, R., & Paninski, L. (2014). *Neuronal dynamics: From single neurons to networks and models of cognition* (1st ed.). Cambridge University Press. https://doi.org/10.1017/CBO9781107447615

Ito, H. T. (2008). Frequency-dependent signal transmission and modulation by neuromodulators. *Frontiers in Neuroscience, 2*(2), 138–144. https://doi.org/10.3389/neuro.01.027.2008

Kuramoto, Y. (1984). *Chemical oscillations, waves, and turbulence* (Vol. 19). Springer. https://doi.org/10.1007/978-3-642-69689-3

La Camera, G., Giugliano, M., Senn, W., & Fusi, S. (2008). The response of cortical neurons to in vivo-like input current: Theory and experiment. *Biological Cybernetics*, *99*(4–5), 279–301. https://doi.org/10.1007/s00422-008-0272-7

Mirollo, R. E., & Strogatz, S. H. (1990). Synchronization of pulse-coupled biological oscillators. *SIAM Journal on Applied Mathematics*, *50*(6), 1645–1662. https://doi.org/10.1137/0150098

Ostojic, S. (2014). Two types of asynchronous activity in networks of excitatory and inhibitory spiking neurons. *Nature Neuroscience*, *17*(4), 594–600. https://doi.org/10.1038/nn.3658

Ott, E., & Antonsen, T. M. (2008). Low dimensional behavior of large systems of globally coupled oscillators. *Chaos: An Interdisciplinary Journal of Nonlinear Science*, *18*(3). https://doi.org/10.1063/1.2930766

Peña Ramirez, J., Olvera, L. A., Nijmeijer, H., & Alvarez, J. (2016). The sympathy of two pendulum clocks: Beyond Huygens' observations. *Scientific Reports*, *6*(1), 23580. https://doi.org/10.1038/srep23580

Pikovsky, A., Rosenblum, M., & Kurths, J. (2001). Synchronization: A universal concept in nonlinear sciences. In *Cambridge nonlinear science series*. Cambridge University Press. https://doi.org/10.1017/CBO9780511755743

Proskurnikov, A. V., & Tempo, R. (2018). A tutorial on modeling and analysis of dynamic social networks. Part II. *Annual Reviews in Control*, *45*, 166–190. https://doi.org/10.1016/j.arcontrol.2018.03.005

Richardson, M. J. E., Brunel, N., & Hakim, V. (2003). From subthreshold to firing-rate resonance. *Journal of Neurophysiology*, *89*(5), 2538–2554. https://doi.org/10.1152/jn.00955.2002

Tao, L., Weber, K. E., Arai, K., & Eden, U. T. (2018). A common goodness-of-fit framework for neural population models using marked point process time-rescaling. *Journal of Computational Neuroscience*, *45*(2), 147–162. https://doi.org/10.1007/s10827-018-0698-4

Uhlhaas, P. J., & Singer, W. (2012). Neuronal dynamics and neuropsychiatric disorders: Toward a translational paradigm for dysfunctional large-scale networks. *Neuron*, *75*(6), 963–980. https://doi.org/10.1016/j.neuron.2012.09.004

Ullner, E., & Politi, A. (2016). Self-sustained irregular activity in an ensemble of neural oscillators. *Physical Review X*, *6*(1), 011015. https://doi.org/10.1103/PhysRevX.6.011015

Ullner, E., Politi, A., & Torcini, A. (2018). Ubiquity of collective irregular dynamics in balanced networks of spiking neurons. *Chaos*, *28*(8). https://doi.org/10.1063/1.5049902

Vera, J., Pereira, U., Reynaert, B., Bacigalupo, J., & Sanhueza, M. (2020). Modulation of frequency preference in heterogeneous populations of theta-resonant neurons. *Neuroscience*, *426*, 13–32. https://doi.org/10.1016/j.neuroscience.2019.10.054

Vreeswijk, C. v., & Sompolinsky, H. (1998). Chaotic balanced state in a model of cortical circuits. *Neural Computation*, *10*(6), 1321–1371. https://doi.org/10.1162/089976698300017214

Wang, M., Li, R., Li, J., Zhang, J., Chen, X., Zeng, S., & Liao, X. (2018). Frequency selectivity of echo responses in the mouse primary auditory cortex. *Scientific Reports*, *8*(1), 49. https://doi.org/10.1038/s41598-017-18465-w

White, J. A., Chow, C. C., Rit, J., Soto-Treviño, C., & Kopell, N. (1998). Synchronization and oscillatory dynamics in heterogeneous, mutually inhibited neurons. *Journal of Computational Neuroscience*, *5*(1), 5–16. https://doi.org/10.1023/A:1008841325921

Yu, S., Yue, W., Guo, T., Liu, Y., Zhang, Y., Khademi, S., Zhou, T., Xu, Z., Song, B., Wu, T., Liu, F., Tai, Y., Yu, X., & Wang, H. (2023). The effect of the sub-threshold oscillation induced by the neurons' resonance upon the electrical stimulation-dependent instability. *Frontiers in Neuroscience*, *17*. https://doi.org/10.3389/fnins.2023.1178606

CHAPTER 4

Dynamical Analysis of a Reaction-Diffusion System for Biocatalytic Exothermic Reactions

'Afifi Md Desa, Mohd Hafiz Mohd
and Mohamad Hekarl Uzir

4.1 INTRODUCTION

Diffusion plays a pivotal role in the functionality of enzymes. The movement of molecules from regions of higher concentration to regions of lower concentration (Mauro, 2021), is essential for the transport of substrates to the active sites of enzymes and the subsequent removal of reaction products. This process underpins many biological systems' efficiency and regulation and is a fundamental concept in biochemistry and molecular biology. According to Segel (1975), specific criteria must be met for substrate molecules and catalytic groups to engage in a reaction, including proximity and alignment. When there is no agitation, the diffusion of the substrate within the enzyme molecule becomes important because the immediate reaction between the substrate and enzyme, which occurs in continuous stirred tank reactors (CSTRs), cannot take place. When studying the diffusion of a substrate into an enzyme molecule, Fick's law of

DOI: 10.1201/9781032678573-4

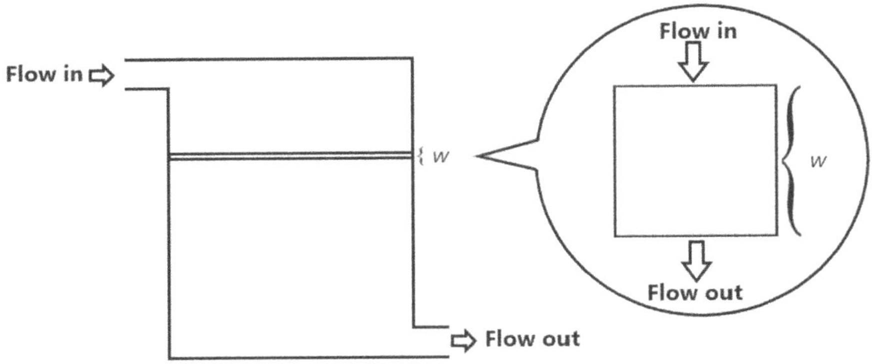

FIGURE 4.1 The substrate enters the reactor, diffusing into the outer surface of the enzyme for the reaction. Substrate diffusion occurs in one direction only, moving from higher to lower concentrations within the enzyme molecule.

diffusion is typically considered. This law describes how diffusion occurs and can be applied to understand the movement of molecules, such as a substrate through a medium (Murray, 2002). Fick's first law, which relates to the steady-state diffusion, states that the diffusion flux is proportional to the concentration gradient. It helps to describe how the concentration of the substrate changes spatially within the enzyme (Mauro, 2021). Meanwhile, Fourier's law of heat conduction describes how heat energy is transferred through a material due to temperature differences. In the context of an enzyme, Fourier's law can be used to understand how temperature changes within the enzyme structure as heat diffuses through it, impacting its activity and stability (Murray, 2002). Figure 4.1 shows the schematic diagram of a continuous tank reactor (CTR). In an unagitated reactor system, substrate introduction initiates diffusion into enzyme molecules. Subsequent chemical transformations occur, leading to the production of the desired product that consequently flows out from the reactor.

4.2 MATHEMATICAL MODELLING

We employ partial differential equations (PDE) to describe the reaction-diffusion system in a biocatalytic reaction. Since it is assumed that the diffusion occurs on the surface of the enzyme, the model is constructed under the following assumptions: (1) the density of the reacting mixture is constant, unaffected by the conversion of substrate to product; (2) the volume of the reacting mixture is constant; (3) no mixing occurs in the reactor, hence the diffusion of substrate A into the surface of enzyme will take place; (4) the specific heat of the reacting mixture remains constant and unaffected by the conversion of substrate to product; (5) the heat of reaction, ΔH is constant, which is independent of temperature and composition; and (6) for simplicity, we assume the diffusion will only occur in one direction. The control volume is shown in Figure 4.1.

By incorporating the substrate and thermal diffusivity into the exothermic biocatalytic reaction system in Desa et al. (2021), the equations to be studied are:

$$\frac{\partial C_A}{\partial t} = \frac{F}{V} C_{A_0} - \frac{F}{V} C_A - \left(\frac{k_A e^{-\frac{E}{RT}} C_A C_{ET}}{C_A + K_M} \right) + \mathscr{D} \frac{\partial^2 C_A}{\partial w^2}$$

$$\frac{\partial T}{\partial t} = \frac{F}{V}(T_0 - T) + \frac{k_A e^{-\frac{E}{RT}} C_A C_{ET}}{\rho C_p (C_A + K_M)} (\Delta H) + \alpha \frac{\partial^2 T}{\partial w^2} - \frac{UA}{\rho V C_p}(T - T_C) \tag{4.1}$$

where F is the feed rate of substrate into the reactor (m³/s), C_{A_0} is the inlet concentration of A (mole/m³), C_A represents the outlet concentration of A, V is the volume of the mixture, k_A is the reaction rate constant, E is the activation energy, R is the universal gas law constant, T_0 is the temperature of the feed stream, C_{ET} is the total enzyme concentration, K_M represents the Michaelis-Menten constant, T is the temperature of the mixture in the reactor, ρ is the density of mixture, C_p is the specific heat of reacting mixture, ΔH is the heat of the reaction, $\mathscr{D}$ is the substrate diffusivity, w is the spatial direction and α is thermal diffusivity.

To simplify the equation and remove the dimensions, we will do the following substitution. Let $y = \frac{C_A}{C_{A_0}}, x = \frac{w\mathscr{D}}{F}$, $\tau = \frac{F}{V}t$, $\theta = \frac{\rho C_p T}{C_{A_0}(\Delta H)}$, $\theta_0 = \frac{\rho C_p T_0}{C_{A_0}(\Delta H)}$ and $T = \frac{\theta C_{A_0}(\Delta H)}{\rho C_p}$. Based on Aris and Amundson (1958), a form of control heat-removal function for $\frac{Q(T)}{F C_{A_0}(\Delta H)}$ was chosen which is given by $q(\theta) = U(\theta - \theta_C)[1 + K_C(\theta - \theta_S)]$. Then followed by $D = \frac{F}{V}$, $K = \frac{K_M}{C_{A_0}}$, $\varepsilon = \frac{C_{ET}}{C_{A_0}}$, $\beta = \frac{E\rho C_P}{R C_{A_0}(\Delta H)}$, $\gamma = \frac{V\mathscr{D}^3}{F^3}$ and

$\phi = \frac{\alpha \mathscr{D}^2 V}{F^3}$ as the dimensionless groups. Upon substituting the terms, the final system of equations reduces into

$$\begin{aligned} \frac{\partial y}{\partial \tau} &= 1 - y - \frac{k_A y \varepsilon e^{-\frac{\beta}{\theta}}}{D(y+K)} + \gamma \frac{\partial^2 y}{\partial x^2} \\ \frac{\partial \theta}{\partial \tau} &= \theta_0 - \theta + \frac{k_A y \varepsilon e^{-\frac{\beta}{\theta}}}{D(y+K)} - U(\theta - \theta_C)\left[1 + K_C(\theta - \theta_S)\right] + \phi \frac{\partial^2 \theta}{\partial x^2} \end{aligned} \tag{4.2}$$

4.3 NUMERICAL METHODS

Neumann boundary conditions are used as the flux of the substrate and the heat flux through the boundary layer are assumed to be equal to the corresponding fluxes entering the surface of the boundary. This is known as zero-flux boundary conditions, given by

$$\left.\frac{\partial y}{\partial x}\right|_{x=0,1} = \left.\frac{\partial \theta}{dx}\right|_{x=0,1} = 0. \tag{4.3}$$

Eq. (4.2), together with the zero-flux boundary conditions in Eq. (4.3), is solved numerically by utilising MATLAB ODE solver, *ode15s*, in one spatial variable x and time t using the method of lines. We also examine the stability of the steady state. In order to accomplish this, the temporal derivative in Eq. (4.2) is set to zero, and the spatial derivative is replaced by a second central difference approximation. The zero-flux boundary conditions as in Eq. (4.3) are approximated with a finite difference approximation in the equations for end points. Using MATLAB *fsolve*, this system is solved for steady state using an initial guess that is identical to the initial condition shown in each figure section. The Jacobian matrix is numerically computed in *fsolve*, and the eigenvalues are then determined using the *eig* function. All eigenvalues must have negative real parts for the steady state to be stable. In addition, we utilised the numerical continuation software XPPAUT to validate our simulation results where stable and unstable steady states along with the bifurcation points are tracked as a model parameter changes.

4.3.1 Surface Plot Analysis

Table 4.1 lists the variables and parameter values used in the numerical method in order to solve the reaction-diffusion system. These values are obtained from Radzi and Uzir (2009) and the value of dimensionless

TABLE 4.1 The Variables and Parameters of the Reaction-Diffusion System: (a) Values Given by Radzi and Uzir (2009) and (b) Diffusivity Constants Given by Swarts et al. (2010) and Deshmukh et al. (2014)

Variable/Parameter	Description	Value
k_A	Reaction velocity constant	e^{25}
K	Dimensionless Michaelis-Menten constant	1
D	Dilution rate: the rate that the existing medium in the reactor is replaced by a fresh medium	3
β	Dimensionless parameter for heat transfer	50
ε	Dimensionless parameter for enzyme concentration	5
θ_0	Dimensionless initial reactor temperature on the stability of the reaction	1.75
θ_C	Dimensionless mean temperature of water in the cooling coil	1.75
θ_S	Desired steady state temperature	2
U	Dimensionless analogue of U_0A, the overall heat transfer rate	1
K_C	Proportional control on the cooling water flow rate constant	$1 \leq K_c \leq 12$

(a)

Variable/Parameter	Description	Value
γ	Dimensionless substrate diffusivity	0.001
ϕ	Dimensionless thermal diffusivity constant	0.1

(b)

substrate diffusivity, γ, and thermal diffusivity, ϕ, are obtained from Swarts et al. (2010) and Deshmukh et al. (2014), respectively.

We initiate the computation of the system's steady state by substituting the parameter values outlined in Table 4.1. Initially, all parameters remain constant, except for the proportional control constant K_C. Through variations in K_C, four distinct dynamic behaviours emerge: (1) bistability between two stable steady states; (2) bistability between a stable steady state and a stable limit cycle; (3) a stable steady state; and (4) stable limit cycles. These diverse dynamics are illustrated in Figures 4.2 to 4.5, facilitating comprehension of the long-term behaviours of the solution. Moreover, these dynamics serve to validate the codimension-one bifurcation diagrams, which are discussed in greater detail in the subsequent section.

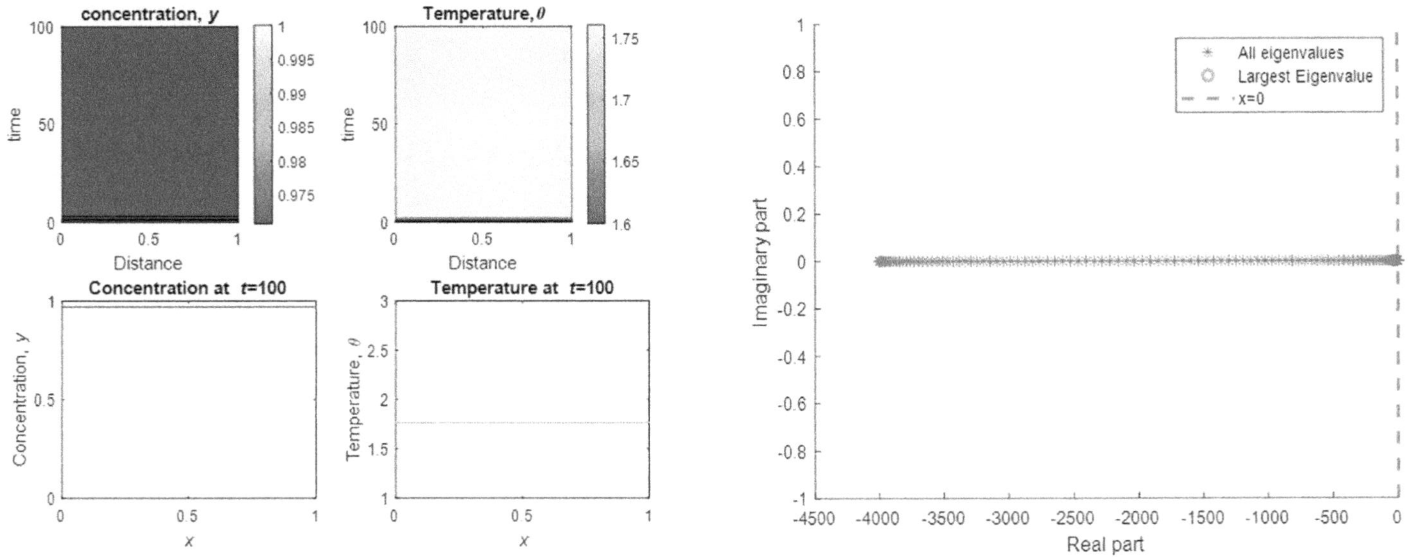

FIGURE 4.2 Surface plot of the system at parameters $K_C = 0$, with the initial condition set at $y = 1, \theta = 1.6$. The plot shows that the steady state is stable with all the real parts of the eigenvalues are negative.

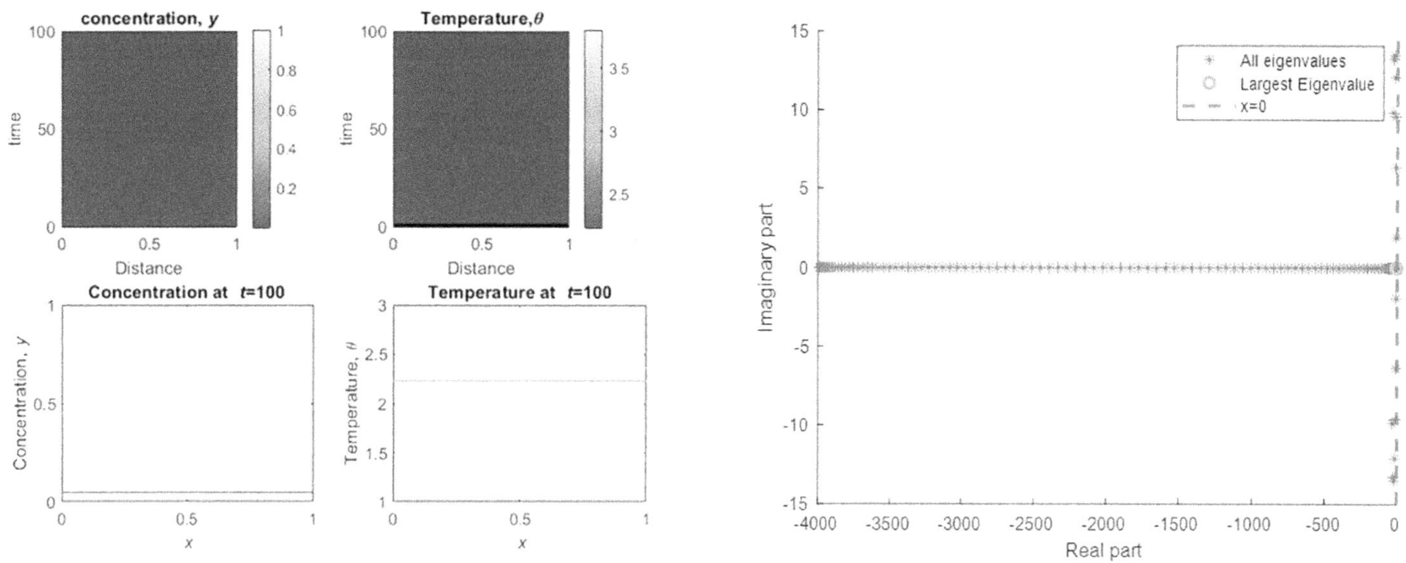

FIGURE 4.3 Surface plot of the system at parameters $K_C = 0$, with the initial condition set at $y = 1, \theta = 3$. The plot shows that the steady state is stable and all the real parts of the eigenvalues are negative.

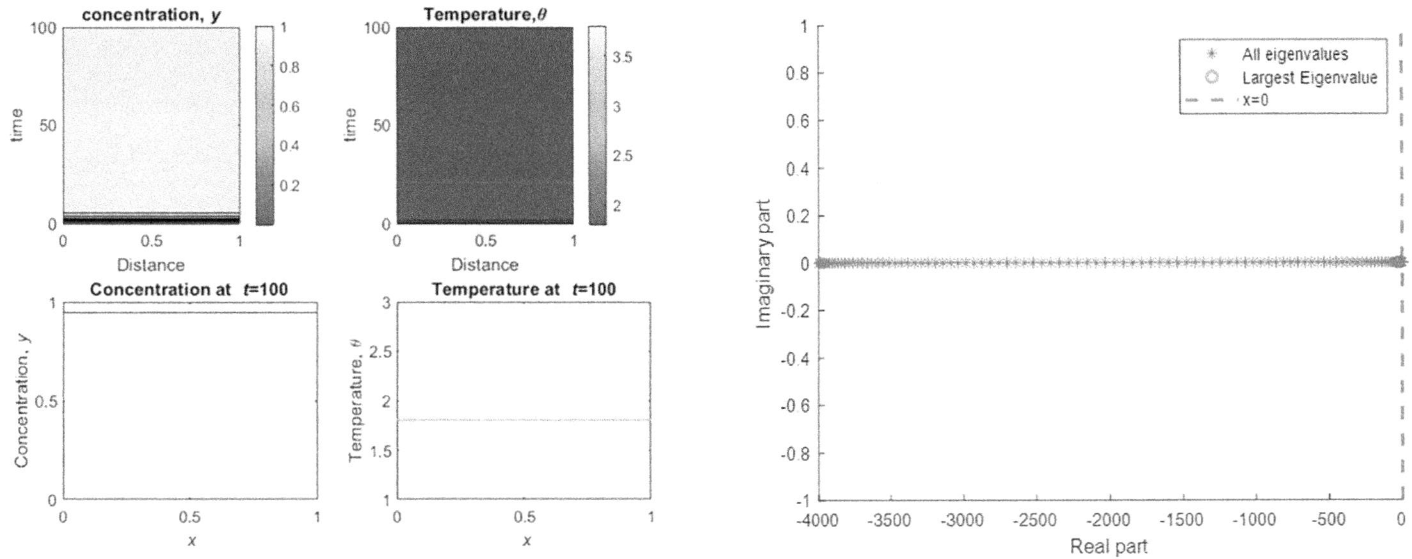

FIGURE 4.4 Surface plot of the system at parameters $K_C = 3$, with the initial condition set at $y = 1, \theta = 1.6$. The plot shows that the steady state is stable with all the real parts of the eigenvalues are negative.

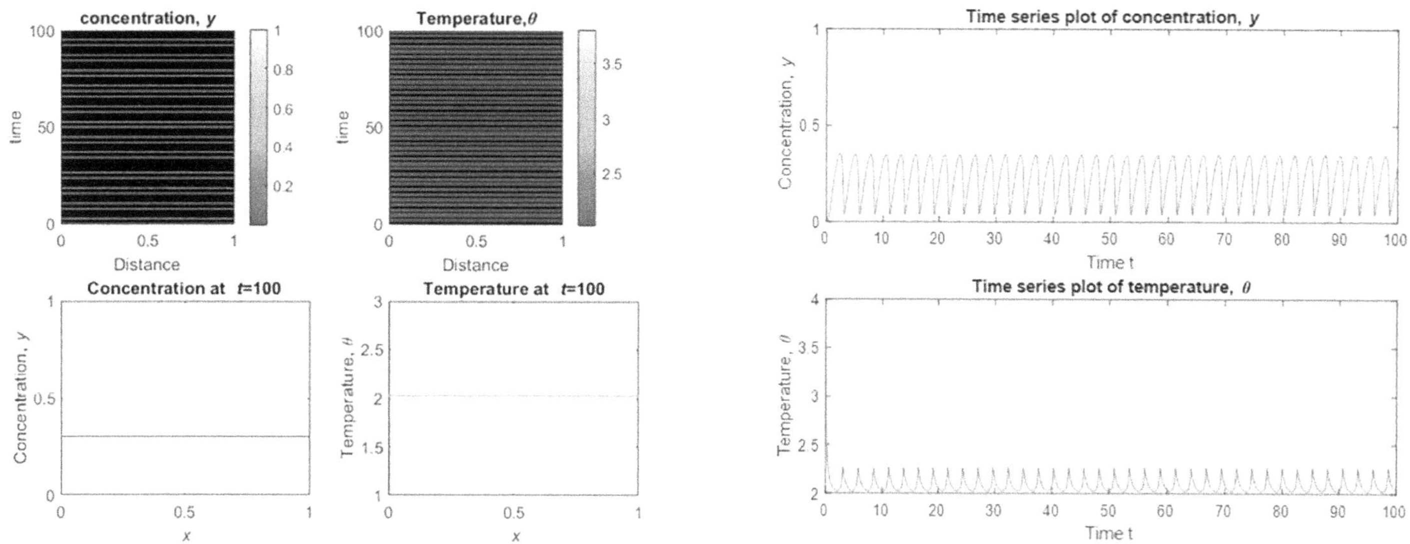

FIGURE 4.5 Surface plot of the system at parameters $K_C = 3$, with the initial condition set at $y = 1, \theta = 3$. The plot shows that the oscillatory solution occurs in the xt-plane. The slice of time series plot reveals the repetitive pattern in concentration and temperature.

Figures 4.2 and 4.3 show the surface plots of the system at parameter $K_C = 0$. We can see that there are two steady states that emerged in the plot. In the absence of temperature control, the system exhibits the emergence of two distinct steady states, depending upon their initial conditions. Both steady states demonstrate stability, supported by the eigenvalue spectrum plotted alongside the surface plot. Notably, all the real parts of the eigenvalues are negative, affirming the stability of both steady states. This phenomenon is also known as bistability between the two stable steady states.

In Figure 4.2, the solution remains stable under high substrate concentrations at lower temperatures. Conversely, in Figure 4.3, the stability is observed at lower substrate concentrations but at higher temperatures.

Figures 4.4 and 4.5 show the existence of another bistability phenomenon, which is the bistability between the stable equilibrium and stable limit cycle. This phenomenon can be observed for the parameter value, $K_C = 3$. The stability of steady state is confirmed through the analysis of spectrum of eigenvalues, revealing that all real parts of the eigenvalues are negative, a characteristic indicative of stable attractors. Notably, the convergence towards these attractors is determined solely by the initial conditions of the system.

As shown in Figure 4.4, when the initial conditions are set at $y = 1$ and $\theta = 1.6$, the solution of the system stabilises at higher substrate concentrations despite a lower reactor temperature. This stability is evident from the time series plot, where both substrate concentration and temperature reach equilibrium. Analysis of the eigenvalues further confirms this equilibrium as a stable node. While in Figure 4.5, when the initial conditions are set at $y = 1$ and $\theta = 3$, the surface plot displays a distinct pattern characterised by repetitive bars across the xt -plane. Upon plotting the time series, we observe oscillations in the solution, particularly at lower substrate concentrations despite a higher reactor temperature. The repetitive pattern in the time series plot suggests that the system returns to similar states over multiple cycles. Interestingly, when different initial conditions are employed, which lead to a minor perturbation, the system returns to the original cycle after the disturbance. This steady state is subsequently confirmed to be a limit cycle in the bifurcation analysis presented in the next section.

Figure 4.6 shows the surface plot of the system at $K_C = 5$. With the increment in K_C, a new dynamical behaviour is observed. The system exhibits a single steady state characterised by higher substrate concentrations with a lower reactor temperature, regardless of the positive initial

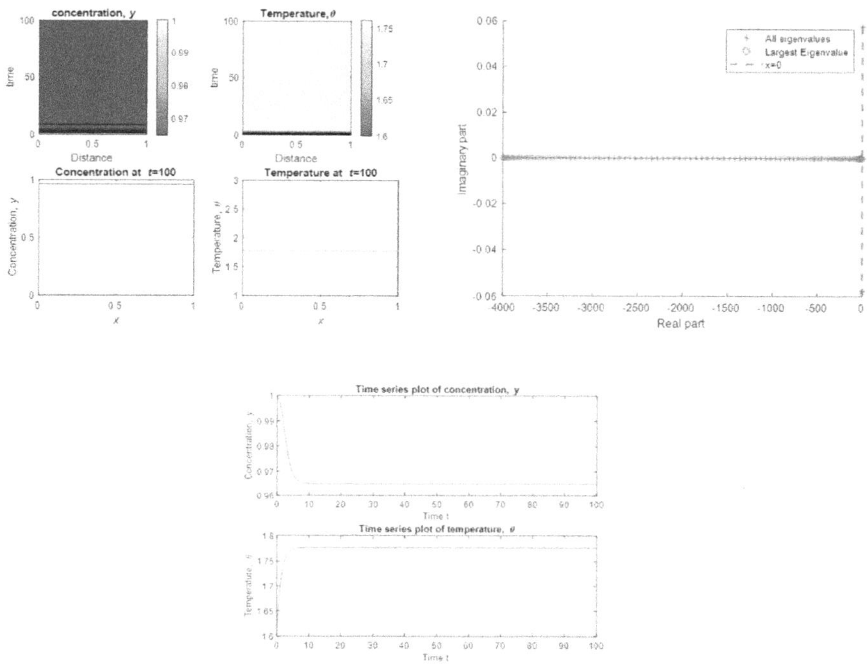

FIGURE 4.6 Surface plot of the system at parameters, $K_C = 5$ for all initial conditions. The plot shows that the steady state is stable and all the real parts of the eigenvalues are negative.

conditions. This equilibrium is validated as a stable node through eigenvalue analysis, where all eigenvalues are real and negative.

Finally, Figure 4.7 shows the surface plot when the parameter is set to $K_C = 9$. The surface plot exhibits a distinct pattern characterised by repetitive bars across the xt -plane, resembling the dynamical behaviour observed in Figure 4.5. However, in this scenario, instead of bistability between a stable limit cycle and a stable steady state, the stable limit cycle emerges as the sole attractor in the system. The time series plot reveals a repetitive pattern, indicating oscillations in both substrate concentration and temperature. This pattern indicates that the system repeatedly returns to similar states across multiple cycles. When different initial conditions are employed, resulting in minor perturbations, they tend to realign the system towards the cycle rather than driving it away.

The analysis of the reaction-diffusion system has revealed a variety of dynamical behaviours, including stable equilibria, bistability phenomenon

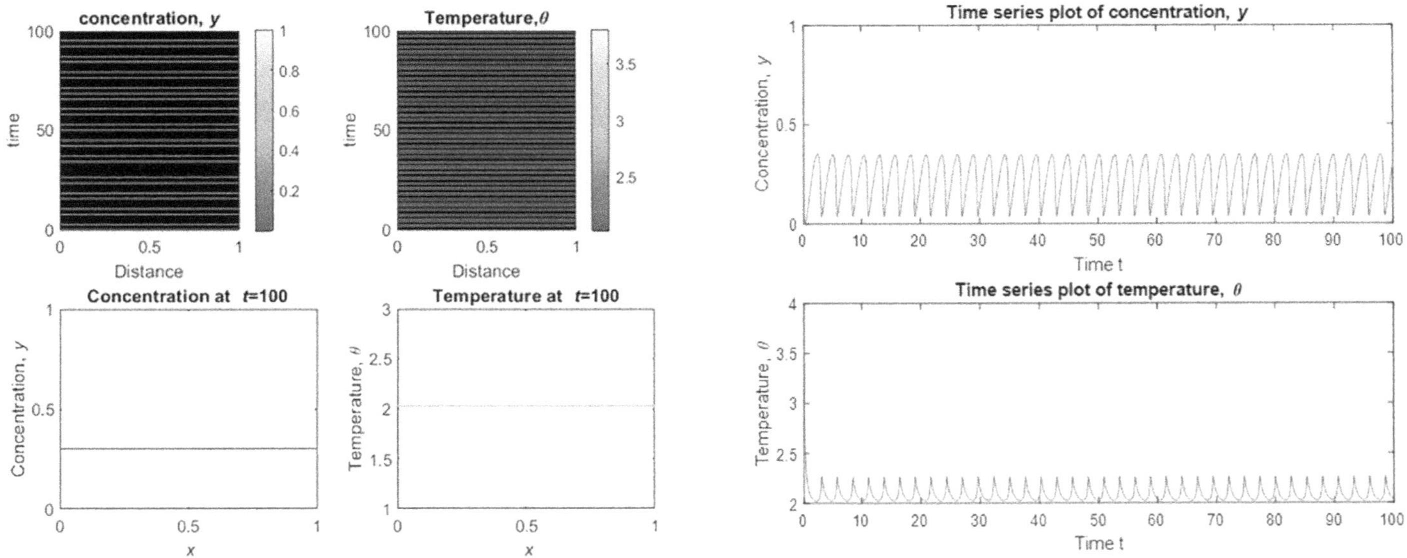

FIGURE 4.7 Surface plot of the system at parameters, $K_C = 9$ for all initial conditions. The plot shows that the oscillatory solution occurs in the *xt*-plane. The slice of time series plot reveals the repetitive pattern in concentration and temperature.

between equilibria and stable limit cycles. These findings have significant implications in understanding the overall behaviour of the systems under different parametrisation and initial conditions. However, to gain a deeper insight into the behaviour of the system and its potential transitions between these attractors, bifurcation analysis for reaction-diffusion system is performed. Bifurcation analysis is essential because it allows us to explore the changes in system's parameters which then lead to qualitative changes in the system's behaviour. By systematically varying key parameters and conducting bifurcation analysis, we can uncover the critical thresholds and transitions within the system, providing valuable insights into its underlying mechanisms and guiding potential applications particularly in control strategies.

4.4 ONE-PARAMETER BIFURCATION ANALYSIS

To investigate the underlying mechanism contributing to the distinct observations in dynamical behaviour of this reaction-diffusion system, we employ a co-dimension one bifurcation analysis using the AUTO feature in XPPAUT. We systematically manipulate the proportional control constant, K_C. This examination involves varying these parameters while monitoring the behaviour of both stable and unstable attractors, as well as identifying points of bifurcation. The bifurcation diagrams are constructed to showcase the effect of varying K_C. In these diagrams, red curves represent stable steady states, black curves represent unstable steady states and green curves indicate the occurrence of stable limit cycles. Abbreviations H, HC and SN correspond to Hopf, homoclinic and saddle-node bifurcations, respectively.

Figure 4.8 shows a bifurcation diagram with K_C serving as the bifurcation parameter. These plots were generated through numerical continuation using XPPAUT. These settings were selected to simulate a scenario characterised by moderate dilution rates and enzyme concentrations. The objective was to effectively illustrate the dynamical changes in the system as K_C was systematically varied from 0 to 12.

It was observed that when the value of $K_C = 0$, the system has three branches of steady state where the upper and lower branches are stable. The stable steady states are separated by an unstable steady state, in the middle branch. This situation leads to the bistability between the two stable steady state on the zero K_C values. This observation is correlated with our finding shown in Figures 4.2 and 4.3. When the value of K_C is increased, there is dynamical change occurring at critical point H_1 on the branch of

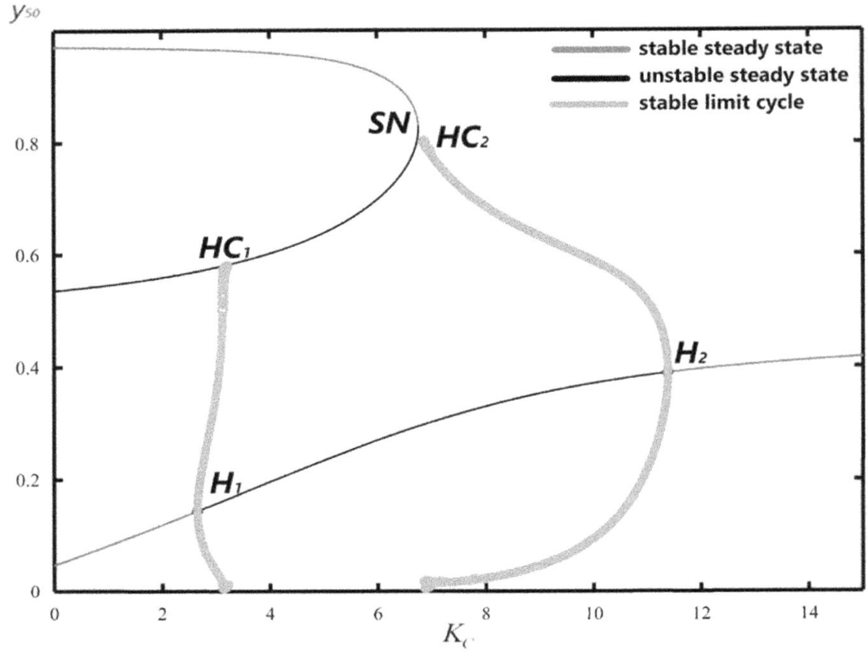

FIGURE 4.8 Bifurcation diagram with K_C as the bifurcation parameter. The value of K_C was systematically varied from 0 to 12. Red curves represent stable steady states, black curves represent unstable steady states and green curves indicate the occurrence of stable limit cycles.

steady states E_1 corresponding to the Hopf bifurcation. The Hopf bifurcation occurs as the emergence of a periodic solution from a previously stable equilibrium. In this situation, the substrate concentration (reactor temperature respectively) starts to oscillate between certain minimum and maximum values. There is also a change in dynamical behaviour as in the previous bistability between the two stable steady states in the region of $K_C < H_1$ is replaced by a bistability between the stable steady states and stable limit cycle in the region of $H_1 < K_C < HC_1$. The surface plot of this bistability phenomenon is shown in Figures 4.4 and 4.5.

As K_C gradually increases, the amplitude of oscillations in the stable limit cycle are also increased and subsequently collide with the unstable steady state in the middle branch at critical point HC_1. This collision leads to a homoclinic bifurcation, culminating in the disappearance of the preceding stable limit cycle via collision with the unstable saddle point.

There is another homoclinic bifurcation occurs at the critical point HC_2 as the result of a collision between the stable limit cycle and unstable saddle point. Between $HC_1 < K_C < HC_2$, a new dynamical behaviour is observed. In this region, there is only a single attractor originating from the upper branch of the stable steady state, which corresponds to the surface plot analysis shown in Figure 4.6. As K_C increases past HC_2, we observe another appearance of a stable limit cycle. This limit cycle results from the second Hopf bifurcation which occurs at the critical point H_2 in the lower branch. This stable limit cycle is the only attractor in the region $HC_2 < K_C < H_2$, which we already showed in Figure 4.7.

Besides that, close to the critical point HC_2, another bifurcation occurs as a result of collision between the stable steady state in the upper branch and the unstable steady state in the middle branch. This collision leads to the saddle node bifurcation, SN. When $K_C > H_2$, the dynamical behaviour changes into a stable single steady state. Only a single attractor exists in the system, which corresponded to a stable steady state on the lower branch.

The effects on increasing K_C are different on each branch. The initial conditions play a vital role in determining the outcome of the system. For instance, at the lower branch of the steady state, increasing the K_C would cause the substrate concentration y to increase. However, the opposite effect is observed on the upper branch of the steady state where increasing K_C would decrease the substrate concentration. When $K_C > SN$, the initial conditions do not affect the system since there is only one steady state existing in the system.

4.5 DISCUSSION

In order to assess the dynamical behaviour of the system, we conducted numerous simulations, testing the effect of changing the proportional control constant on the outcome of the system. Interestingly, we observe different distinct dynamical behaviours that occurred in the system.

Building on these findings, we conduct a bifurcation analysis to investigate the effects of the proportional control constant. Notably, four distinct dynamic behaviours have been identified: (1) bistability between two stable steady states, (2) bistability between a stable steady state and a stable limit cycle, (3) a stable steady state and (4) stable limit cycles.

In a biocatalytic reaction, understanding the dynamics by identifying the critical transitions between potential dynamical regimes is essential. Key parameters, such as temperature, are vital in achieving high reactor efficiency and productivity. From the bifurcation diagram in Figure 4.8, it is evident that the region of bistability exists at very lower values of the proportional control constant e.g., $K_C < 2$. Bistability is characterised by the coexistence of two stable steady states, depending on the initial conditions (Ferrell, 2002). In control systems, this phenomenon is generally undesirable, as a single, stable steady state is preferred for ensuring consistent and reliable performance. Control systems typically strive to achieve a single stable equilibrium for predictability and dependability (Khalil, 2002). Furthermore, at intermediate values, in the range of $4 < K_C < 10$, the system exhibits a stable limit cycle, which is another undesirable behaviour as denoted by Sontag (2008); oscillatory dynamics can disrupt system predictability and stability, making a stable steady state more desirable. To optimise the control system performance and operability, the proportional control constant should be either in the lower range, such as $2 < K_C < 4$, or in higher values, such as $K_C > 10$.

The results from our analysis highlight the significant influence of temperature on the dynamics of exothermic biocatalytic reactions, especially within a real reactor setting. In industrial reactors, temperature fluctuations can profoundly affect the reaction rates and equilibrium, often leading to non-linear behaviour such as thermal runaway or oscillatory patterns (Ball & Gray, 2013). This is due to the exothermic nature of the reactions, where heat generation accelerates reaction rates, potentially causing localised hotspots that can lead to enzyme denaturation and reactor instability (Chang et al., 2019; Schwolow et al., 2016). The interplay between reaction heat and diffusion is critical: efficient heat removal systems must be implemented to prevent excessive temperature build-up, which could shift the system towards undesirable states as indicated by our bifurcation analysis. Such findings are consistent with previous studies (Akaje & Olajuwon, 2021; Chatterjee & Chaturvedi, 2011; Han et al., 2007), where inadequate temperature control in reactors has been shown to result in reduced yield and selectivity of desired products.

Enzymes, being proteinaceous in nature, are highly sensitive to temperature variations, which directly impacts their catalytic efficiency and stability (Li et al., 2022). Our analysis suggests that even a slight temperature increase can lead to significant changes in enzyme activity, potentially reducing reaction rates or leading to enzyme inactivation if the temperature exceeds the enzyme's optimal range. This sensitivity is particularly pronounced in biocatalytic systems, where the activation energy for substrate conversion is relatively low, making the system highly responsive to temperature changes (Potkule et al., 2024). It is well documented that enzymes exhibit a bell-shaped activity profile with respect to temperature, with a sharp decline in activity beyond a certain threshold due to thermal denaturation (Elias et al., 2014; Gault et al., 2021). Therefore maintaining an optimal temperature range within the reactor is crucial to ensure sustained enzyme activity and overall process efficiency.

Overall, this study enhances our understanding of the reaction-diffusion system and its stability in biocatalytic systems, thus providing valuable insights for process optimisation and design in reactors.

4.6 CONCLUSION

This work provides valuable insights into the behaviour of a reaction-diffusion system in biocatalysis. The findings emphasise that diffusion significantly affects the dynamical behaviour of the system, leading to the existence of oscillatory states, alterations in stability patterns and the phenomenon of bistability.

Practically, this implies that in real production scenarios, where diffusion plays a crucial role, the system may not perform as efficiently as an idealised continuous stirred tank reactor system in a perfectly mixed reactor. These insights align with previous research (Chävez & Kapral, 2001; Hess & Plesser, 1979; Santos et al., 2006) highlighting the impact of diffusion on reaction rates and conversion efficiencies.

Overall, this study enhances our understanding of the reaction-diffusion system and their stability in biocatalytic systems, thus providing valuable insights for process optimisation and design in reactors.

4.7 ACKNOWLEDGEMENT

The authors acknowledge support from the Fundamental Research Grant Scheme with Project Code: FRGS/1/2022/STG06/USM/02/1 by the Ministry of Higher Education, Malaysia (MOHE).

REFERENCES

Akaje, T. W., & Olajuwon, B. I. (2021). Thermal diffusion and diffusion thermo effects on Casson nanofluid flow in the presence of chemical reaction with inclined magnetic field. *Asian Research Journal of Mathematics*, *16*(12), 110–124. https://doi.org/10.9734/arjom/2020/v16i1230259

Aris, R., & Amundson, N. R. (1958). An analysis of chemical reactor stability and control—II: The evolution of proportional control. *Chemical Engineering Science*, *7*(3), 132–147. https://doi.org/10.1016/0009-2509(58)80020-2

Ball, R., & Gray, B. F. (2013). Thermal instability and runaway criteria: The dangers of disregarding dynamics. *Process Safety and Environmental Protection*, *91*(3), 221–226. https://doi.org/10.1016/j.psep.2012.05.008

Chang, Y.-H., Kishore, J., & Shadman, F. (2019). Run-away energetic reactions in the exhaust of deposition reactors. *Advances in Chemical Engineering and Science*, *09*(02), 223–238. https://doi.org/10.4236/aces.2019.92017

Chatterjee, A., & Chaturvedi, S. (2011). Heat transfer in a reaction-diffusion system with a moving heat source. *International Journal of Heat and Mass Transfer*, *54*(1–3), 326–337. https://doi.org/10.1016/j.ijheatmasstransfer.2010.09.039

Chävez, F., & Kapral, R. (2001). Compartmentalized reaction-diffusion systems. *Physical Review E—Statistical, Nonlinear, and Soft Matter Physics*, *63*(1 II), 1–9. https://doi.org/10.1103/PhysRevE.63.016211

Desa, A. M., Mohd, M. H., & Uzir, M. H. (2021). Bifurcation analysis of an exothermic biocatalytic reaction system. *Pertanika Journal of Science & Technology*, *29*(1).

Deshmukh, P. S., Manjunatha, S. S., & Raju, P. S. (2014). Influence of moisture content on thermophysical properties of enzyme clarified sapota (Achras sapota L.) juice. *International Food Research Journal*, *21*(4), 1539–1552.

Elias, M., Wieczorek, G., Rosenne, S., & Tawfik, D. S. (2014). The universality of enzymatic rate-temperature dependency. *Trends in Biochemical Sciences*, *39*(1), 1–7. https://doi.org/10.1016/j.tibs.2013.11.001

Ferrell, J. E. (2002). Self-perpetuating states in signal transduction: Positive feedback, double-negative feedback and bistability. *Current Opinion in Cell Biology*, *14*(2), 140–148. https://doi.org/10.1016/S0955-0674(02)00314-9

Gault, S., Higgins, P. M., Cockell, C. S., & Gillies, K. (2021). A meta-analysis of the activity, stability, and mutational characteristics of temperature-adapted enzymes. *Bioscience Reports*, *41*(4), 1–10. https://doi.org/10.1042/BSR20210336

Han, Y. M., Li, Z., & Luo, J. L. (2007). Temperature waves in chemical reaction-diffusion-heat conduction systems with two ends respectively subject to Dirichlet and no-flux conditions. *Chinese Chemical Letters*, *18*(11), 1427–1430. https://doi.org/10.1016/j.cclet.2007.09.021

Hess, B., & Plesser, T. (1979). Temporal and spatial order in biochemical systems. *Annals of the New York Academy of Sciences*, *316*(1), 203–213. https://doi.org/10.1111/j.1749-6632.1979.tb29470.x

Khalil, H. K. (2002). *Nonlinear systems* (3rd ed.). CDSWEBCERNCH. https://doi.org/10.1016/j.physa.2006.08.011

Li, G., Buric, F., Zrimec, J., Viknander, S., Nielsen, J., Zelezniak, A., & Engqvist, M. K. M. (2022). Learning deep representations of enzyme thermal adaptation. *Protein Science*, *31*(12), 1–25. https://doi.org/10.1002/pro.4480

Mauro, J. C. (2021). Fick's laws of diffusion. *Materials Kinetics*, 39–58. https://doi.org/10.1016/B978-0-12-823907-0.00026-1

Murray, J. D. (2002). Mathematical biology: I. An introduction. In *Interdisciplinary applied mathematics* (Vol. 1, Issue 1, 3rd ed.). https://doi.org/10.1086/421587

Potkule, J. B., Kahar, S. P., Kumar, M., & Annapure, U. S. (2024). Impact of non-thermal techniques on enzyme modifications for their applications in food.

International Journal of Biological Macromolecules, 275(P1), 133566. https://doi.org/10.1016/j.ijbiomac.2024.133566

Radzi, M. R. M., & Uzir, M. H. (2009). Stability study of an exothermic biocatalytic reaction and its application in bioprocess systems. *Pertanika Journal of Science and Technology, 17*(June 2007), 95–115.

Santos, J. dos, Lozano, R., Friboulet, A., & Mondie, S. (2006). Prediction of unstable behavior in enzymatic diffusion-reaction Processes, 2082–2087. In *Proceedings of the 45th IEEE conference on decision and control* (pp. 2082–2087). San Diego, CA, USA. https://doi.org/10.1109/CDC.2006.376927

Schwolow, S., Ko, J. Y., Kockmann, N., & Röder, T. (2016). Enhanced heat transfer by exothermic reactions in laminar flow capillary reactors. *Chemical Engineering Science, 141*, 356–362. https://doi.org/10.1016/j.ces.2015.11.022

Segel, I. H. (1975). *Enzyme kinetics: Behavior and analysis of rapid equilibrium and steady state enzyme systems.* Wiley.

Sontag, E. D. (2008). Network reconstruction based on steady-state data. *Essays in Biochemistry, 45*, 161–176. https://doi.org/10.1042/BSE0450161

Swarts, J. W., Kolfschoten, R. C., Jansen, M. C. A. A., Janssen, A. E. M., & Boom, R. M. (2010). Effect of diffusion on enzyme activity in a microreactor. *Chemical Engineering Journal, 162*(1), 301–306. https://doi.org/10.1016/j.cej.2010.04.040

CHAPTER 5

Stability Analysis of Magnetohydrodynamic Flow and Heat Transfer over an Exponentially Permeable Stretching/Shrinking Sheet in Ferrofluids with the Influence of Heat Generation

Norshafira Ramli, Nurul Syuhada Ismail and Syakila Ahmad

5.1 INTRODUCTION

The first successful colloid was created in the early 1960s by a researcher named Papell (1965) at NASA, despite the fact that the scientific world had long dreamed of and struggled to produce a liquid with strong magnetic

DOI: 10.1201/9781032678573-5

properties. Ferrofluid is a colloidal suspension of magnetic nanoparticles dispersed in a non-conducting base fluid, with the particles ranging in diameter from 5 to 15 nm (Sheikholeslami & Rashidi, 2015). Research on ferrofluid is being conducted at several eminent institutions worldwide. This substance has drawn interest from a number of researchers and provides difficult basic research opportunities in small particle magnetism and dispersion chemistry. Many engineering fields, including mechanical, electromechanical, electrical, biological, chemical and acoustical, have looked into and explored its properties for practical uses (Raj & Chorney, 1998).

Meanwhile, there has been a lot of interest in the use of ferrofluid, which is a strongly magnetised liquid exposed to a magnetic field that is used as a coolant to improve heat transfer in heat exchangers and electrical devices on both small and large scales. When an external magnetic field is present, the ferrofluid's thermal conductivity parameters are noticeably enhanced (Bahiraei & Hangi, 2013). The excellence of the issue is demonstrated by the numerous studies that have been reported recently on the influence of ferrofluid applied to various problems under an external magnetic field, which increases thermal performance. Lajvardi et al. (2010) examined the ferrofluid forced convection heat transfer in the presence of a steady magnetic field, and their research produced a discernible improvement in heat transfer. The primary cause of the heat transfer coefficient enhancement may be attributed to the significant modifications in the thermophysical characteristics of the ferrofluid generated by the applied magnetic field. A circular copper tube filled with ferrofluid and subjected to a fluctuating magnetic field while maintaining laminar flow and homogeneous heat flux was the subject of an experimental investigation by Ghofrani et al. (2013). The largest improvement in convection heat transfer, observed at a low Reynolds number (Re), was 27.6%. An experimental investigation on the laminar forced-convection heat transfer of ferrofluids in a parallel plate heated uniformly under an external magnetic field generated by four electromagnets was carried out by Goharkhah et al. (2016). They found that, at a constant Re value, the magnetic field intensity raises the heat transfer rate by convective heat transfer, which is directly correlated with both Re value and nanoparticle concentration. Subsequently, the issues surrounding ferrofluid have garnered significant interest, and a substantial body of literature has been discovered in several contexts, including Ghasemian et al. (2015), Bezaatpour and Rostamzadeh (2020), Ramli and Ahmad (2021), Mat Yasin et al. (2022), Ebaid et al. (2022), Sun et al. (2023) and Mehrez (2024).

Research on boundary layer flows caused by a stretching/shrinking sheet has significant real-world applications in the metallurgy and chemical industries, including aerodynamic plastic sheet extrusion, the manufacturing of glass fibre, the drying and cooling of paper and textiles and other related applications (Roşca et al., 2016). Studies on linear or nonlinear stretching/shrinking flat sheets have dominated research over the past few decades, with the study of boundary layer flow over exponentially stretching/shrinking sheets receiving less attention. Research on the flow across an exponentially stretching continuous surface appears to have begun with Magyari and Keller (1999). In their work, the boundary layers' properties related to mass and heat transmission were explained and contrasted with the findings of other writers, who got their conclusions using the more well-known power-law boundary conditions. Then, Bachok et al. (2012) explored the two-dimensional stagnation-point flow of a water-based nanofluid over an exponentially stretching/shrinking sheet. Studying an exponentially stretching/shrinking sheet shows that, in comparison to a linear stretching/shrinking instance, the range of the stretching/shrinking parameter where the similarity solution occurs is wider. Rohni et al. (2012) examined the subject of constant laminar two-dimensional flow and heat transmission over an exponentially shrinking vertical sheet with suction. A scrutiny of flow and heat transfer on stagnation point over an exponentially stretching/shrinking cylinder filled with nanofluid in the presence of slip was examined by Najib et al. (2022). They observed that the heat transfer rate at the surface was immediately increased by the region of dual solutions expanding in the presence of slip and curvature factors, while the rate of heat transfer at the surface rapidly dropped when the cylinder shrunk. Various studies on exponentially stretching/shrinking sheets can also be found in the works of Hafidzuddin et al. (2016), Anuar et al. (2019), Mohd Azmi et al. (2020), Yashkun et al. (2021), Allaw et al. (2023) and Zainal et al. (2023).

In addition, while controlling the efficiency of the heat transfer, heat generation and absorption also become crucial factors. Mahmoud and Waheed (2012) considered a problem on magnetohydrodynamic (MHD) flow and heat transfer of a micropolar fluid with heat generation/absorption and slip effect. They confirmed that, as the heat generation parameter is raised, the local Nusselt number decreases. After that, MHD stagnation flow of nanofluid past a stretching/shrinking surface with the influence of slip and heat generation/absorption is considered by Nandy and

Mahapatra (2013) by using the shooting method. Jusoh et al. (2019) investigated the details of stagnation point flow in nanofluid with heat generation and concluded that the process of heat transfer was hampered by heat generation, while it was enhanced by heat absorption. Later, Wahid et al. (2021) elucidated the effect of heat generation on the hybrid nanofluid slip flow over an exponentially stretching/shrinking sheet, and they found that increases in the copper volume percentage and heat generation parameters can improve the temperature profile; however, increases in the velocity, thermal slips and suction parameters can have the opposite effect. Also, in the study on hybrid nanofluid, Eid and Nafe (2020) also took the slip effect and the existence of heat generation into account. They claimed that the temperature rises as the heat generation parameter values rise. Furthermore, the effect of heat generation/absorption on MHD along a stretching cylinder with a porous medium is contemplated by Reddy et al. (2023).

Nonetheless, research on stability analysis on MHD flow and heat transfer of ferrofluid over an exponentially stretching/shrinking sheet with the influence of heat generation has not been documented in the literature. The goal of this work is to extend the MHD flow and heat transfer over an exponentially permeable stretching/shrinking sheet in ferrofluids, as described by Rasli and Ramli (2021). The controlling partial differential equations are reduced to a set of nonlinear ordinary differential equations by appropriate similarity transformations. With the use of the bvp4c function in MATLAB, the resulting equations are numerically solved. A detailed analysis is conducted, and the graphical representations that provide information and discussion of the influences on several parameters over significant physical quantities are unveiled. The stability analysis is carried out for the non-unique solutions that are evident and the smallest eigenvalues that imply the stability of the flow are tabulated.

5.2 BASIC EQUATIONS

Consider a steady, laminar and two-dimensional flow of an incompressible ferrofluids and heat transfer over an exponentially permeable stretching/shrinking sheet in the presence of a transverse magnetic field with heat generation effect, as presented in Figure 5.1. Assuming that the surface temperature at the wall varies exponentially in the form $T_w = T_\infty + T_0 e^{x/2L}$, where T_∞ and T_0 represent the free stream temperature assumed to be constant and reference temperature, respectively, and

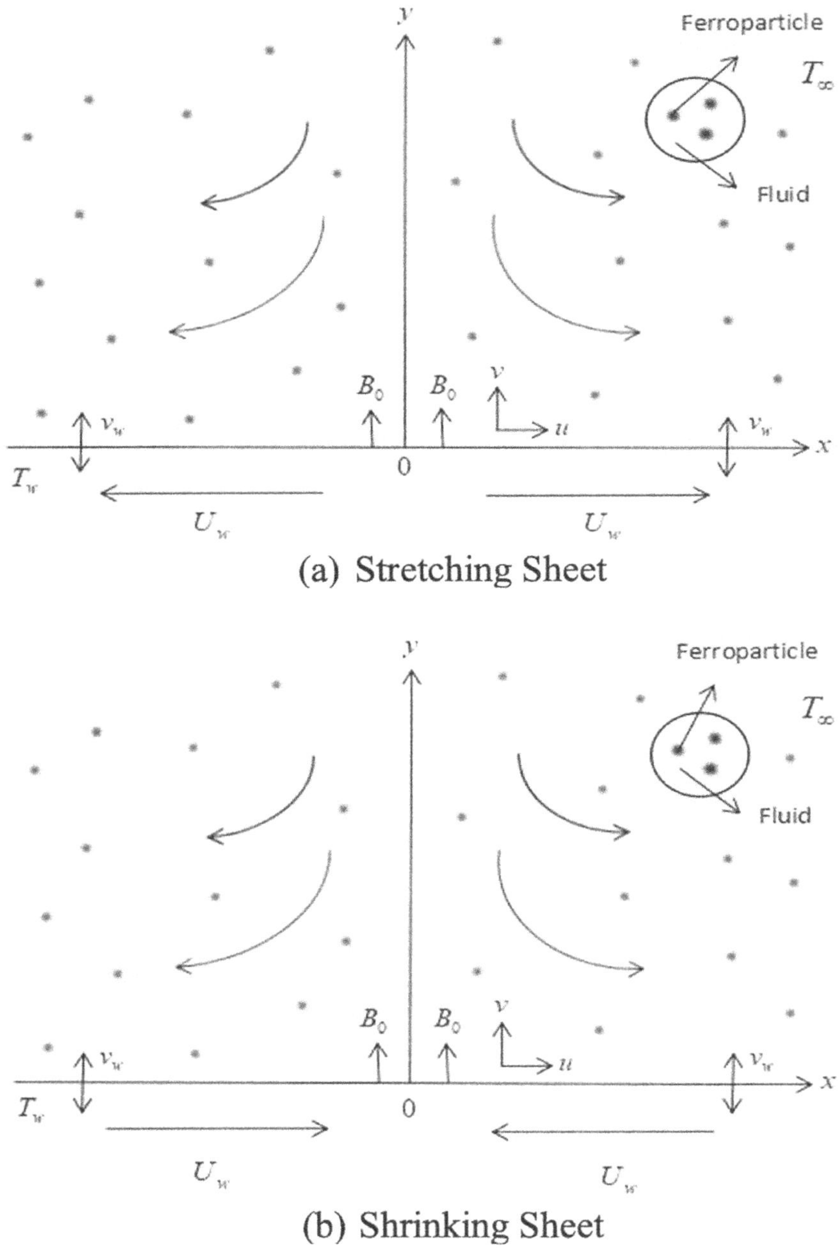

FIGURE 5.1 A schematic representation of flow geometry.

L denotes the distinctive length. The governing equations that regulate the ferrofluid in the free stream under consideration, are written as follows (see Bhattacharyya, 2011; Rasli & Ramli, 2021):

$$\frac{\partial u}{\partial x}+\frac{\partial v}{\partial y}=0, \tag{5.1}$$

$$u\frac{\partial u}{\partial x}+v\frac{\partial u}{\partial y}=\frac{\mu_{ff}}{\rho_{ff}}\frac{\partial^2 u}{\partial y^2}-\frac{\sigma_{ff}B^2}{\rho_{ff}}u, \tag{5.2}$$

$$u\frac{\partial T}{\partial x}+v\frac{\partial T}{\partial y}=\frac{k_{ff}}{\left(\rho C_p\right)_{ff}}\frac{\partial^2 T}{\partial y^2}+\frac{q}{\left(\rho C_p\right)_{ff}}\left(T-T_\infty\right), \tag{5.3}$$

subject to the boundary conditions:

$$u=\lambda U_w,\quad v=v_w,\quad T=T_w\quad \text{at}\quad y=0,$$
$$u\to 0,\quad T\to T_\infty\quad \text{as}\quad y\to\infty. \tag{5.4}$$

where u and v are the velocity components along the x- and y-axes, respectively, B is the total magnetic field, T is the fluid temperature in the boundary layer, q is the dimensional heat generation, λ is the constant stretching/shrinking parameter, U_w is the stretching/shrinking velocity and v_w is the variable wall mass transfer velocity. Here, μ_{ff} is the viscosity of the ferrofluid, ρ_{ff} is the density of the ferrofluid, σ_{ff} is an electrical conductivity of the ferrofluid, k_{ff} is the thermal conductivity of the ferrofluid and $\left(\rho C_p\right)_{ff}$ is the heat capacity of the ferrofluid, which are defined as follows:

$$\mu_{ff}=\frac{\mu_f}{\left(1-\varphi\right)^{2.5}},\quad \alpha_{ff}=\frac{k_{ff}}{\left(\rho C_p\right)_{ff}},\quad \rho_{ff}=\left(1-\varphi\right)\rho_f+\varphi\rho_s,$$

$$\left(\rho C_p\right)_{ff}=\left(1-\varphi\right)\left(\rho C_p\right)_f+\varphi\left(\rho C_p\right)_s,$$

$$\frac{\sigma_{ff}}{\sigma_f}=1+\frac{3\left(\frac{\sigma_s}{\sigma_f}-1\right)\varphi}{\left(\frac{\sigma_s}{\sigma_f}+2\right)-\left(\frac{\sigma_s}{\sigma_f}-1\right)\varphi},\quad \frac{k_{ff}}{k_f}=\frac{k_s+2k_f-2\varphi\left(k_f-k_s\right)}{k_s+2k_f+\varphi\left(k_f-k_s\right)}, \tag{5.5}$$

where μ_f is the viscosity of the base fluid, φ is the ferroparticle volume fraction, α_{ff} is the thermal diffusivity of the ferrofluid, ρ_f and ρ_s are the densities of the base fluid and the solid, $\left(\rho C_p\right)_f$ and $\left(\rho C_p\right)_s$ are the heat capacities of the base fluid and solid, σ_f and σ_s are the electrical conductivity of the base fluid and solid, and k_f and k_s are the thermal conductivities of the base fluid and of the solid, respectively. The variable magnetic field B is taken in the form of

$$B = B_0 e^{x/2L}, \tag{5.6}$$

where B_0 is a constant magnetic field. The dimensional heat generation q equates to

$$q = q_0 e^{x/L}, \tag{5.7}$$

where q_0 is the heat generation constant. The stretching/shrinking velocity U_w is given by

$$U_w = a e^{x/L}, \tag{5.8}$$

where a is the stretching/shrinking velocity rate with $a > 0$ for stretching and $a < 0$ for shrinking. Then, the variable wall mass transfer velocity v_w in the boundary conditions in Eq. (5.4) is given by

$$v_w = v_0 e^{x/2L} \tag{5.9}$$

where v_0 is a constant with $v_0 < 0$ for mass suction and $v_0 > 0$ for mass injection.

The governing Eqs. (5.1), (5.2) and (5.3) subject to the boundary conditions in Eq. (5.4) can be expressed in a simpler form by introducing the following similarity transformation:

$$\eta = y\left(\frac{a}{2\nu_f L}\right)^{1/2} e^{x/2L}, \quad \psi = \left(2\nu_f La\right)^{1/2} f(\eta) e^{x/2L}, \quad \theta(\eta) = \frac{T - T_\infty}{T_w - T_\infty}, \tag{5.10}$$

where η is the similarity variable, ν_f is the kinematic viscosity of the base fluid and ψ is the stream function defined as $u = \partial\psi / \partial y$ and $v = -\partial\psi / \partial x$. Thus, we have

$$u = a e^{x/L} f'(\eta), \quad v = -\sqrt{\frac{\nu_f a}{2L}} e^{x/2L} \left[f(\eta) + \eta f'(\eta)\right]. \tag{5.11}$$

Employing the similarity variables in Eq. (5.10), Eqs. (5.2) and (5.3) reduce to the following nonlinear ordinary differential equations:

$$\frac{1}{(1-\varphi)^{2.5}\left[(1-\varphi)+\varphi\rho_s/\rho_f\right]}f'''-2f'^2+ff''$$
$$-\frac{\sigma_{ff}}{\sigma_f\left[(1-\varphi)+\varphi\rho_s/\rho_f\right]}M\,f'=0, \tag{5.12}$$

$$\frac{1}{\Pr}\frac{k_{ff}}{k_f}\theta''+\left[1-\varphi+\varphi\left(\rho C_p\right)_s/\left(\rho C_p\right)_f\right]\left(f\theta'-f'\theta\right)+\beta\theta=0, \tag{5.13}$$

where primes denote differentiation with respect to η, $M=2L\sigma_f B_0^2/\rho_f a$ is the magnetic parameter, $\Pr=\left(\mu C_p\right)_f/k_f$ is the Prandtl number and $\beta=2q_0L/a\left(\rho C_p\right)_f$ is the heat generation parameter.

The boundary conditions in Eq. (5.4) also reduce to the following form:

$$f(0)=S,\quad f'(0)=\lambda,\quad \theta(0)=1,$$
$$f'(\eta)\to 0,\quad \theta(\eta)\to 0\quad \text{as}\quad \eta\to\infty, \tag{5.14}$$

where $S=-v_0\sqrt{2L/\nu_f a}$ is the wall mass transfer parameter, where $S>0$ corresponds to mass suction and $S<0$ corresponds to mass injection. Moreover, λ is the stretching/shrinking parameter with $\lambda>0$ for a stretching sheet and $\lambda<0$ for a shrinking sheet.

The physical quantities of interest are the skin friction coefficient C_f and the local Nusselt number Nu_x, which are respectively defined as

$$C_f=\frac{\tau_w}{\rho_f U_w^2},\quad Nu_x=\frac{Lq_w}{k_f\left(T_w-T_\infty\right)}, \tag{5.15}$$

where the surface shear stress τ_w and the surface heat flux q_w are given by

$$\tau_w=\mu_{ff}\left(\frac{\partial u}{\partial y}\right)_{y=0},\quad q_w=-k_{ff}\left(\frac{\partial T}{\partial y}\right)_{y=0}. \tag{5.16}$$

Using the similarity variables in Eq. (5.10), we obtain

$$\left(2/\mathrm{Re}\right)^{1/2}C_f=\frac{1}{(1-\varphi)^{2.5}}f''(0), \tag{5.17}$$

$$\left(2/\text{Re}\right)^{1/2} Nu_x = -\frac{k_{ff}}{k_f}\theta'(0), \tag{5.18}$$

where $\text{Re} = U_w L/\nu_f$ is the local Reynolds number.

5.3 STABILITY ANALYSIS

In this study, we are using the stability analysis developed by Weidman et al. (2006). According to Weidman and co-workers, a variable τ has to be introduced. In order to perform the stability analysis, unsteady cases are considered. Hence, Eqs. (5.2) and (5.3) are replaced by

$$\frac{\partial u}{\partial t} + u\frac{\partial u}{\partial x} + v\frac{\partial u}{\partial y} = \frac{\mu_{ff}}{\rho_{ff}}\frac{\partial^2 u}{\partial y^2} - \frac{\sigma_{ff}B^2}{\rho_{ff}}u, \tag{5.19}$$

$$\frac{\partial T}{\partial t} + u\frac{\partial T}{\partial x} + v\frac{\partial T}{\partial y} = \frac{k_{ff}}{\left(\rho C_p\right)_{ff}}\frac{\partial^2 T}{\partial y^2} + \frac{q}{\left(\rho C_p\right)_{ff}}\left(T - T_\infty\right), \tag{5.20}$$

where t refers to the time. Then, the new dimensionless variables are unveiled for the unsteady problem based on the variables given by Eq. (5.10) previously used, such that

$$\eta = y\left(\frac{a}{2\nu_f L}\right)^{1/2} e^{x/2L}, \quad \psi = \left(2\nu_f La\right)^{1/2} f(\eta,\tau)e^{x/2L},$$
$$\theta(\eta) = \frac{T - T_\infty}{T_w - T_\infty}, \quad \tau = \frac{at}{2L}e^{x/L}. \tag{5.21}$$

Consequently, Eqs. (5.19) and (5.20) are modified to the following equations:

$$\frac{1}{(1-\varphi)^{2.5}\left[(1-\varphi)+\varphi\rho_s/\rho_f\right]}\frac{\partial^3 f}{\partial\eta^3} - 2\left(\frac{\partial f}{\partial\eta}\right)^2 + f\frac{\partial^2 f}{\partial\eta^2}$$
$$-\frac{\sigma_{ff}}{\sigma_f}\frac{M}{1-\varphi+\varphi\rho_s/\rho_f}\frac{\partial f}{\partial\eta} - \left(1+2\tau\frac{\partial f}{\partial\eta}\right)\frac{\partial^2 f}{\partial\eta\,\partial\tau} + 2\tau\frac{\partial f}{\partial\tau}\frac{\partial^2 f}{\partial\eta^2} = 0, \tag{5.22}$$

$$\frac{k_{ff}/k_f}{1-\varphi+\varphi\left(\rho C_p\right)_s/\left(\rho C_p\right)_f}\frac{1}{\Pr}\frac{\partial^2\theta}{\partial\eta^2} + f\frac{\partial\theta}{\partial\eta} - \frac{\partial f}{\partial\eta}\theta$$
$$+\frac{\beta}{1-\varphi+\varphi\left(\rho C_p\right)_s/\left(\rho C_p\right)_f}\theta - \left(1+2\tau\frac{\partial f}{\partial\eta}\right)\frac{\partial\theta}{\partial\tau} + 2\tau\frac{\partial f}{\partial\tau}\frac{\partial\theta}{\partial\eta} = 0, \tag{5.23}$$

with the following boundary conditions

$$f(0,\tau)=S,\ \ \frac{\partial f}{\partial \eta}(0,\tau)=\lambda,\ \ \theta(0,\tau)=1,$$

$$\frac{\partial f}{\partial \eta}(\eta,\tau)\to 0,\ \ \theta(\eta,\tau)\to 0. \tag{5.24}$$

In order to examine the stability of the steady flow solution $f(\eta)=f_0(\eta)$ and $\theta(\eta)=\theta_0(\eta)$ that satisfy the boundary-value problem in Eqs. (5.12) to (5.14), we introduce

$$f(\eta,\tau)=f_0(\eta)+e^{-\gamma\tau}F(\eta,\tau),\theta(\eta,\tau)=\theta_0(\eta)+e^{-\gamma\tau}G(\eta,\tau). \tag{5.25}$$

It is noteworthy to mention that γ is an unknown eigenvalue and $F(\eta,\tau)$ and $G(\eta,\tau)$ are small relative to $f_0(\eta)$ and $\theta_0(\eta)$. Later on, Eq. (5.25) is substituted into Eqs. (5.22) and (5.23) to produce the following equations

$$\begin{aligned}&\frac{1}{(1-\varphi)^{2.5}\left[(1-\varphi)+\varphi\rho_s/\rho_f\right]}\frac{\partial^3 F}{\partial \eta^3}+f_0\frac{\partial^2 F}{\partial \eta^2}\\&+\left(\gamma-4f_0'-\frac{\sigma_{ff}}{\sigma_f}\frac{M}{1-\varphi+\varphi\rho_s/\rho_f}+2\tau f_0'\gamma\right)\frac{\partial F}{\partial \eta}\\&+\left(f_0''-2\tau\gamma f_0''\right)F+2\tau f_0''\frac{\partial F}{\partial \tau}-\left(1+2\tau f_0'\right)\frac{\partial^2 F}{\partial \eta\,\partial \tau}=0,\end{aligned} \tag{5.26}$$

$$\begin{aligned}&\frac{k_{ff}/k_f}{1-\varphi+\varphi(\rho C_p)_s/(\rho C_p)_f}\frac{1}{\Pr}\frac{\partial^2 G}{\partial \eta^2}+f_0\frac{\partial G}{\partial \eta}\\&+\left(\gamma-f_0'+\frac{\beta}{1-\varphi+\varphi(\rho C_p)_s/(\rho C_p)_f}+2\tau f_0'\gamma\right)G\\&-\frac{\partial F}{\partial \eta}\theta_0+\left(\theta_0'-2\tau\gamma\theta_0'\right)F+2\tau\frac{\partial F}{\partial \tau}\theta_0'-\left(1+2\tau f_0'\right)\frac{\partial G}{\partial \tau}=0,\end{aligned} \tag{5.27}$$

together with the following boundary conditions

$$F(0,\tau)=0,\ \ \frac{\partial F}{\partial \eta}(0,\tau)=0,\ \ G(0,\tau)=0,$$

$$\frac{\partial F}{\partial \eta}(\eta,\tau)\to 0,\quad G(\eta,\tau)\to 0. \tag{5.28}$$

After that, we set $\tau=0$, $F=F_0(\eta)$ and $G=G_0(\eta)$ to obtain the following:

$$\frac{1}{(1-\varphi)^{2.5}\left[(1-\varphi)+\varphi\rho_s/\rho_f\right]}F_0'''+f_0F_0'' \\ +\left(\gamma-4f_0'-\frac{\sigma_{ff}}{\sigma_f}\frac{M}{1-\varphi+\varphi\rho_s/\rho_f}\right)F_0'+f_0''F_0=0, \tag{5.29}$$

$$\frac{k_{ff}/k_f}{1-\varphi+\varphi(\rho C_p)_s/(\rho C_p)_f}\frac{1}{\Pr}G_0''+f_0G_0' \\ +\left(\gamma-f_0'+\frac{\beta}{1-\varphi+\varphi(\rho C_p)_s/(\rho C_p)_f}\right)G_0-F_0'\theta_0+\theta_0'F_0=0, \tag{5.30}$$

and the boundary conditions

$$F_0(0)=0,\quad F_0'(0)=0,\quad G_0(0)=0, \\ F_0'(\eta)\to 0,\quad G_0(\eta)\to 0. \tag{5.31}$$

By relaxing a boundary condition on $G_0(\eta)$ or $F_0'(\eta)$, we can determine the possible values of γ. The objective of the procedure is to ascertain the stability of the problem, and for this current study, we relax $F_0'(\infty)\to 0$. The system was then solved at the boundary $F_0''(0)=1$ using the new condition.

5.4 RESULTS AND DISCUSSION

Using the bvp4c solver in MATLAB, the numerical solutions to Eqs. (5.12) and (5.13) conditioned on Eq. (5.14) are investigated. For the solver's coding function to compute the most accurate results, appropriate initial guesses, the ideal boundary layer thickness and different parameter values must be chosen and carefully modified. This study takes into account surfaces that are stretching ($\lambda>0$) and shrinking ($\lambda<0$), and the effects of other parameters, including the magnetic parameter M, heat generation parameter β and wall mass transfer parameter S, are examined on the heat transfer and boundary layer flow. Throughout the study, the

Prandtl number will remain constant at $\Pr = 6.2$ for water-based fluid and for the volume fraction of solid ferroparticles, the range $0 \le \varphi \le 0.2$ is considered, where $\varphi = 0$ represents pure fluid water.

The thermophysical properties of water-based and ferroparticles are depicted in Table 5.1 (see Ramli & Ahmad, 2021). By making direct comparisons with the values of $f''(0)$ and $-\theta'(0)$ with those obtained by Hafidzuddin et al. (2016) and Wahid et al. (2021), as shown in Table 5.2, the accuracy of the numerical method used is evaluated. A good agreement between the findings is obtained, indicating that the numerical approach used is accurate. Because of this, we have great confidence in the accuracy of our numerical findings. Figures 5.2 to 5.11 show the skin friction coefficient $(2\mathrm{Re})^{1/2} C_f$ and local Nusselt number $(2/\mathrm{Re})^{1/2} Nu_x$, as well as the velocity $f'(\eta)$ and temperature $\theta(\eta)$ profiles for various values of M, S and β. For Eqs. (5.12) to (5.14), it is noticeable that there are dual solutions, first and second solutions throughout the range of $\lambda > \lambda_c$, where λ_c is the critical value of λ. Furthermore, the numerical solutions to those equations did not exist for the case of $\lambda > \lambda_c$.

The behaviour of the $(2\mathrm{Re})^{1/2} C_f$ is shown in Figure 5.2, where M fluctuates from 0 to 1 with respect to λ. There is an increase in $(2\mathrm{Re})^{1/2} C_f$ when $\lambda < 0$ for

TABLE 5.1 Thermophysical Properties of Water-Based and Ferroparticles

Thermophysical Properties	Water-Based	Ferroparticles		
		Fe_3O_4	$CoFe_2O_4$	$Mn\text{-}ZnFe_2O_4$
$C_p\,(\mathrm{J/kg\cdot K})$	4179	670	700	800
$\rho\,(\mathrm{kg/m^3})$	997.1	5180	4907	4900
$k\,(\mathrm{W/m\cdot K})$	0.613	9.7	3.7	5
$\sigma\,(\mathrm{m^2/s})$	5.5×10^{-6}	0.74×10^{6}	3.9×10^{6}	8×10^{6}

TABLE 5.2 Comparison of $f''(0)$ and $-\theta'(0)$ When $\lambda = -1$, $S = 3$, $\Pr = 0.7$ and $M = \beta = \varphi = 0$

		Hafidzuddin et al. (2016)	Wahid et al. (2021)	Present Study
$f''(0)$	First solution	2.3908	2.3908	2.3908
	Second solution	−0.9722	−0.9721	−0.9722
$-\theta'(0)$	First solution	1.7712	1.7712	1.7712
	Second solution	0.8483	0.8477	0.8483

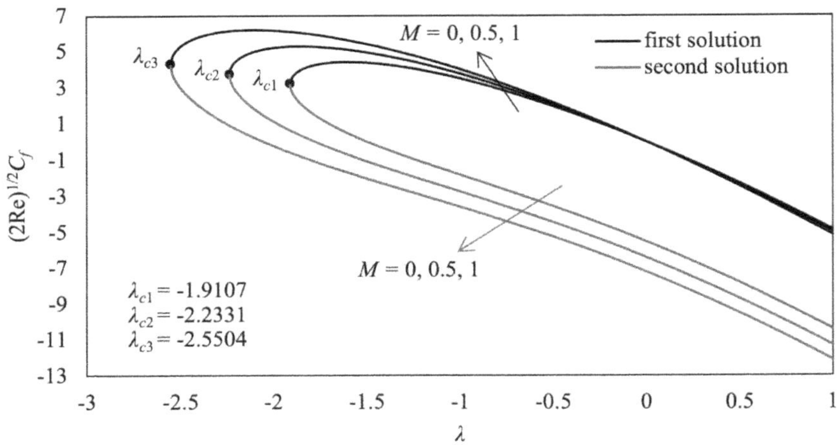

FIGURE 5.2 Variation of $(2\text{Re})^{1/2} C_f$ with λ for several values of M.

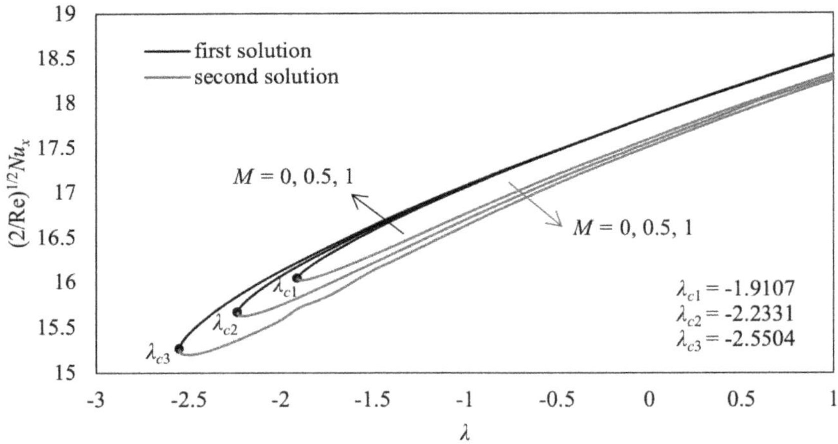

FIGURE 5.3 Variation of $(2/\text{Re})^{1/2} Nu_x$ with λ for several values of M.

the first solution and decrease for the second solution when M increases. A transverse magnetic field applied to an electrically conducting fluid flow experiences a drag force known as the Lorentz force, which acts in the opposite direction of the flow and causes an effect of flow retardation. As a result, expectations regarding the latter qualitatively agree. Figure 5.3 depicts how the local Nusselt number $(2/\text{Re})^{1/2} Nu_x$ behaves in relation to λ, as M changes. The improved heat transfer rate beyond the permeable stretching/shrinking sheet is shown by the first solution with an increase in the values of M. However, the second solution shows the opposite behaviour. More magnetic nanoparticles are drawn to the surface of the stretching/shrinking sheet as the magnetic force gets stronger, which lowers the fluid's thermal conductivity.

Figures 5.4 and 5.5 indicate the $(2\mathrm{Re})^{1/2} C_f$ and $(2/\mathrm{Re})^{1/2} Nu_x$ showing the effect of the mass transfer parameter for suction, $S > 0$. Figure 5.4 shows an increment in S when $\lambda < 0$, which gives an upsurge of $(2\mathrm{Re})^{1/2} C_f$ for the first solution. On the other hand, the second solution gives the decrement in $(2\mathrm{Re})^{1/2} C_f$ as S rises. This figure shows that the flow separation occurs a bit later as the suction power increases. The slow-moving molecules are trapped by the permeable sheet together with the stretching/shrinking sheet, which helps to maintain the laminar flow regime. Consequently, the flow separation is delayed and the wall shear stress at the sheet's surface rises. As portrayed in Figure 5.5, the $(2/\mathrm{Re})^{1/2} Nu_x$ is seen to increase with the enhancement of S for the first and second solutions. The rate of heat transfer increases when suction is present.

The numerical results of $(2/\mathrm{Re})^{1/2} Nu_x$ with heat generation parameter β is depicted in Figure 5.6. It is noticed that it gives nearly the same critical values $\lambda_c = -2.2331$ for the varying values of β. This figure makes it clear that the rate of transfer reduces as heat generation rises, with the presence of suction. Additionally, the rate at which heat is transferred increases

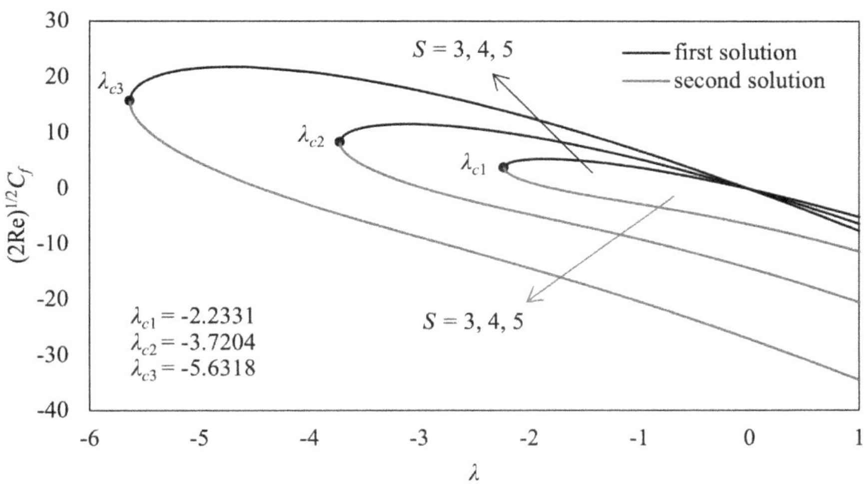

FIGURE 5.4 Variation of $(2\mathrm{Re})^{1/2} C_f$ with λ for several values of S.

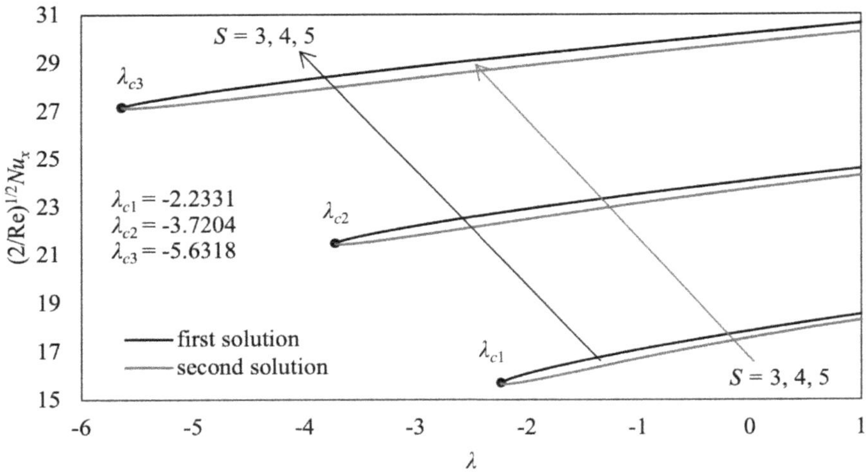

FIGURE 5.5 Variation of $(2/\text{Re})^{1/2} Nu_x$ with λ for several values of S.

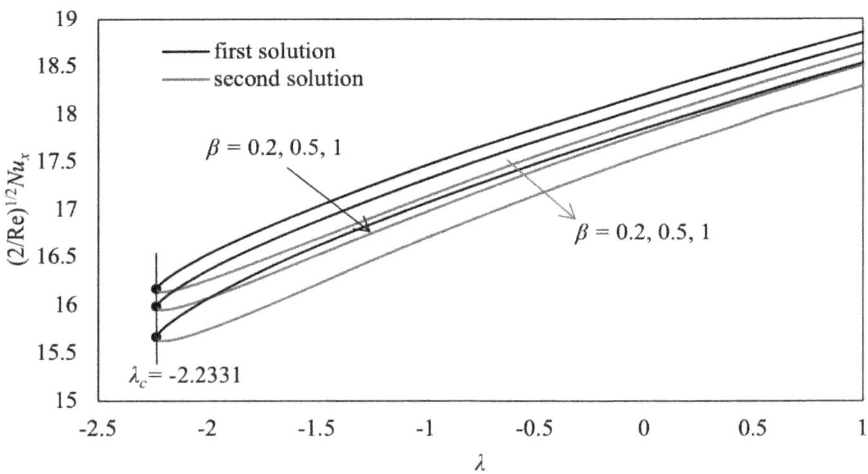

FIGURE 5.6 Variation of $(2/\text{Re})^{1/2} Nu_x$ with λ for several values of β.

as the suction parameter grows but decreases when the heat generation parameter increases. Due to such relationships, increasing the suction parameter causes the transverse fluid motion to accelerate and the rate of heat transfer to increase.

The velocity $f'(\eta)$ and temperature $\theta(\eta)$ profiles are illustrated in Figures 5.7 to 5.11. The numerical findings obtained are validated by these figures, which asymptotically satisfy the far field boundary requirements given in Eq. (5.14). Moreover, it can be observed that the first solution's momentum and thermal boundary layer thicknesses are lower than those of the second solution, indicating the relative stability of the first solution. The effect of M on the temperature and velocity profiles will now be examined. The velocity profile $f'(\eta)$ fluctuation for various values of M in the case of a shrinking sheet is shown in Figure 5.7. The figure shows that for a shrinking sheet, velocity at a given position increases for the first solution as values of M increase and drops for the second solution. This arises physically from the Lorentz force for a ferrofluid given by the last term in Eq. (5.12), which is a result of the imposed magnetic field having a retarding effect on the flow. According to Figure 5.8, for the first solution, temperature falls with increasing M and increases with increasing M for the second solution, which relate directly with the physical phenomena of the ferrofluid system. The stretching/shrinking sheet diminishes as M increases, as a result, the advection of fluid velocity above the sheet affects the temperature field, and can be controlled by a strong external magnetic field.

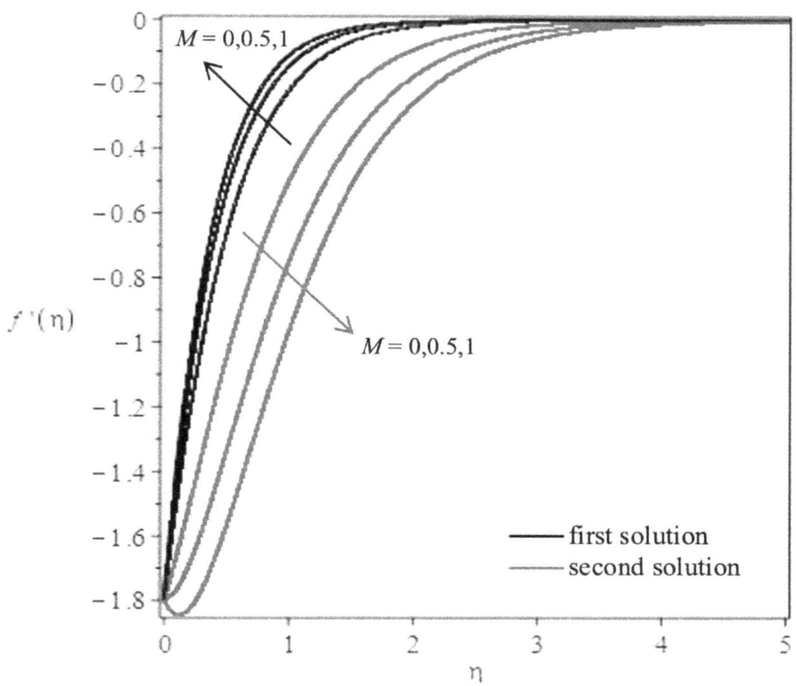

FIGURE 5.7 Velocity profile for various values of M when $\lambda = -1.8$.

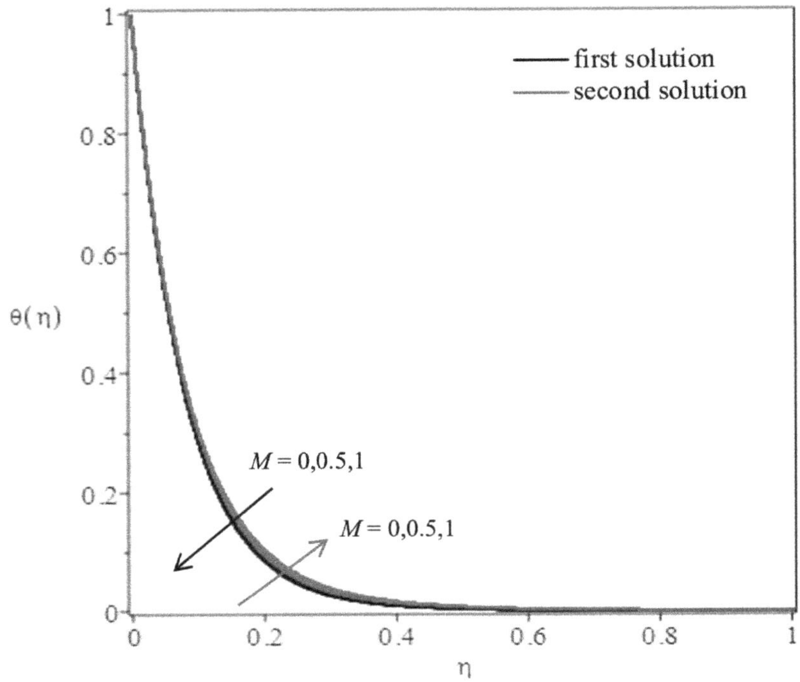

FIGURE 5.8 Temperature profile for various values of M when $\lambda = -1.8$.

Figures 5.9 and 5.10 show velocity and temperature profiles at various values of mass suction parameter, S. The velocity field of the ferrofluid is accelerated as the suction term values are increased, as shown in Figure 5.9. Because of the increase in the colloidal suspension of ferroparticles, which consequently improved the particle dispersion in water-based, the flow magnitude therefore increased. The non-settling nature of the ferrofluid mixture causes free collisions between the ferroparticles, which raises the velocity profile. Moreover, the increment in S causes the temperature profile $\theta(\eta)$ to reduce for both solutions as portrayed in Figure 5.10. A rise in the suction force applied to the surface causes the pressure to drop. The adiabatic cooling effect could cause a drop in temperature as a result of this pressure drop. The fluid tends to lose kinetic energy as it is dragged or driven towards a surface, and this might cause the temperature to drop.

Within Figure 5.11, the influence of heat generation parameter over $\theta(\eta)$ is revealed. It is evident that the presence of heat generation causes the thermal state to increase. This phenomenon is primarily caused by the

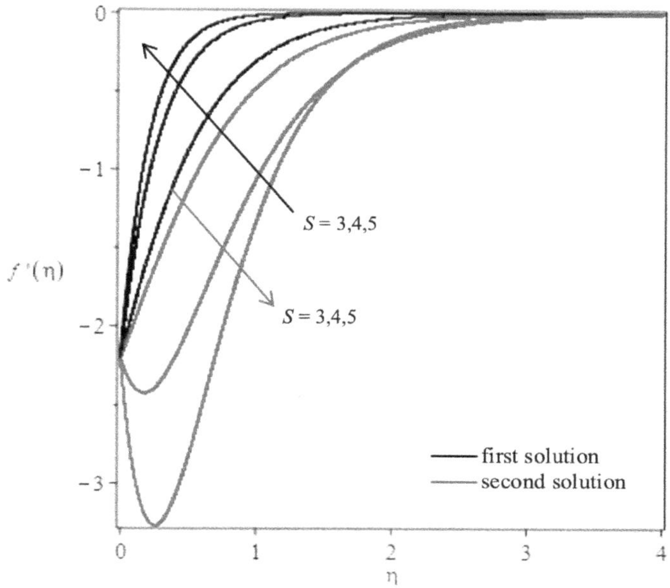

FIGURE 5.9 Velocity profile for various values of S when $\lambda = -2.2$.

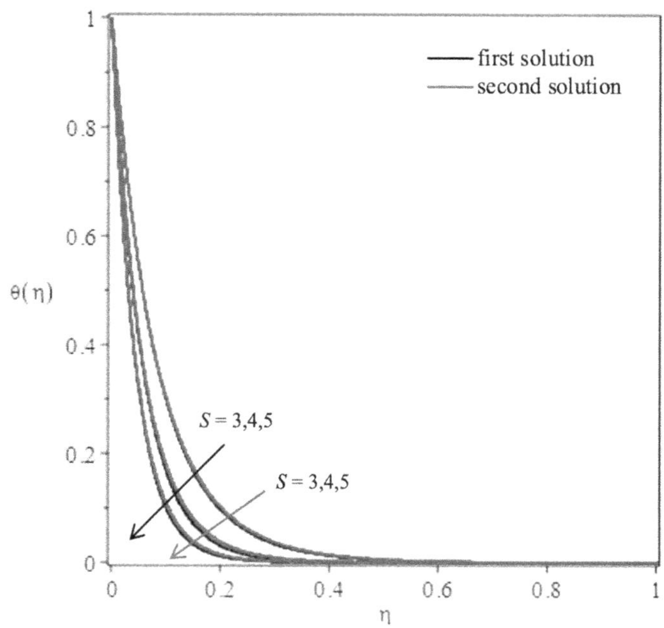

FIGURE 5.10 Temperature profile for various values of S when $\lambda = -2.2$.

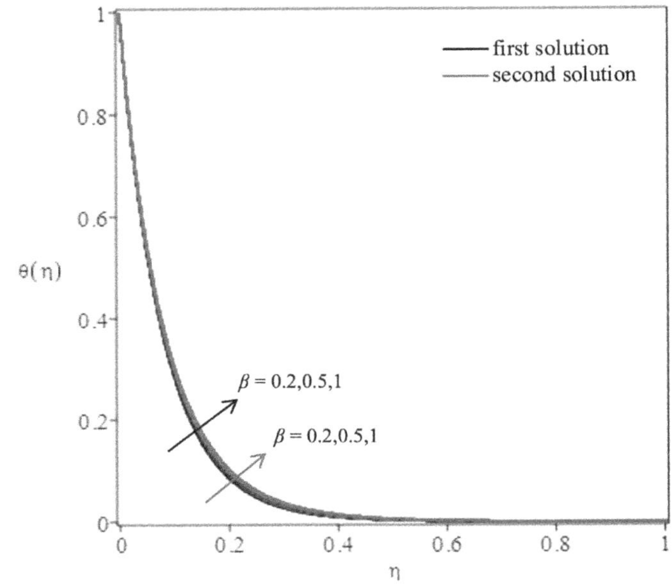

FIGURE 5.11 Temperature profile for various values of β when $\lambda = -2.2$.

extra energy generated in the boundary layer, according to the fluid's heat generation. This extra energy then multiplies the thickness of the thermal boundary layer, which is correlated with temperature and results from the buoyancy effect.

By replacing the set of linearised Eqs. (5.29) and (5.30) with the boundary conditions (5.31), the stability analysis is carried out in MATLAB using the bvp4c solver. The stability of the solutions can be ascertained by identifying the smallest eigenvalue. If the smallest eigenvalue is positive, the flow is stable; if it is negative, the flow is unstable. Table 5.3 shows the smallest eigenvalues for various magnetic and shrinking parameters. It is noteworthy that the smallest eigenvalue approaches 0 when the shrinking rate approaches the critical value, λ_c. Additionally, the positive smallest eigenvalue changed signs to the negative eigenvalue at the turning point of the first and second solutions. The first solutions are discovered to have the positive smallest eigenvalues, while the second solutions have the negative smallest eigenvalues. Thus, we can conclude that the first solution is stable and physically reliable and the second solution is unstable and not physically reliable.

One of the most crucial considerations when attempting to accelerate the rate of heat transfer is probably the type of ferrofluid used. In order to determine which ferroparticle has the highest cooling capability, a comparison of several of them is made. Table 5.4 illustrates the consequences

TABLE 5.3 The Smallest Eigenvalues for Fe_3O_4 Ferroparticle at Several Values of λ with Various of M When $S=3$, $\varphi=0.1$, $\Pr=6.2$ and $\beta=1$

		γ	
M	λ	**First Solution**	**Second Solution**
0	−1.91	0.0729	−0.0728
	−1.88	0.4622	−0.4583
	−1.86	0.5910	−0.5848
	−1.84	0.6944	−0.6859
0.5	−2.23	0.3584	−0.3556
	−2.20	0.5557	−0.5490
	−2.18	0.6846	−0.6763
	−2.14	0.8994	−0.8850
1	−2.55	0.0995	−0.0994
	−2.54	0.3406	−0.3349
	−2.53	0.4751	−0.4658
	−2.52	0.5731	−0.5682

TABLE 5.4 Variation of $(2/\mathrm{Re})^{1/2} Nu_x$ with λ (Stretching and Shrinking Sheet) for Different Ferroparticles at Several Values of β When $M=0.5$, $S=3$, $\varphi=0.1$ and $\Pr=6.2$

			$(2/\mathbf{Re})^{1/2} \boldsymbol{Nu_x}$	
λ	β	**Ferroparticle**	**First Solution**	**Second Solution**
−1	0.2	Fe_3O_4	17.4576	17.1329
		$CoFe_2O_4$	18.5224	18.2269
		Mn-$ZnFe_2O_4$	18.3255	18.0268
	0.5	Fe_3O_4	17.3129	16.9733
		$CoFe_2O_4$	18.3783	18.0703
		Mn-$ZnFe_2O_4$	18.1829	17.8716
	1	Fe_3O_4	17.0658	16.6975
		$CoFe_2O_4$	18.1325	17.7966
		Mn-$ZnFe_2O_4$	17.9398	17.6011
1	0.2	Fe_3O_4	18.8562	18.6351
		$CoFe_2O_4$	19.9394	19.7363
		Mn-$ZnFe_2O_4$	19.7400	19.5352
	0.5	Fe_3O_4	18.7340	18.5064
		$CoFe_2O_4$	19.8169	19.6081
		Mn-$ZnFe_2O_4$	19.6190	19.4086
	1	Fe_3O_4	18.5267	18.2873
		$CoFe_2O_4$	19.6094	19.3905
		Mn-$ZnFe_2O_4$	19.4141	19.1934

of several types of ferrofluids on the local Nusselt number, $(2/\mathrm{Re})^{1/2} Nu_x$ with stretching and shrinking, λ cases at several values of heat generation parameter, β. The outcome manifests that, in contrast to the shrinking case, the stretching case has a larger heat transfer rate. Table 5.1 reveals that $CoFe_2O_4$ has the lowest value of thermal conductivity, k when compared to other ferroparticles. Furthermore, we can observe from this table that the cobalt ferrite ferroparticle $CoFe_2O_4$ gives the highest local Nusselt number compared to other ferroparticles and consequently results in the highest heat transfer rate. Therefore, a decrease in the thermal conductivity of $CoFe_2O_4$ may lead to greater temperature gradients and improved heat transfer results.

5.5 CONCLUSION

In addition to stability study, we have numerically examined the impact of heat generation on MHD flow and heat transfer over an exponentially permeable stretching/shrinking sheet in ferrofluids. In the case of a stretching/shrinking sheet, we can sum up by saying that dual solutions are obtained for ferrofluids. It is discovered that raising the magnetic parameter value increases the ferrofluid's range of solution, skin friction coefficient and heat transfer rate. Additionally, the skin friction coefficient and heat transfer rate give an increment as the mass transfer parameter (suction) increases. The enhancement of heat transfer rate can be perceived as the heat generation parameter drops. It has been shown that for ferrofluids, the cobalt ferrite ferroparticle in the water-based ferrofluid has the maximum heat transfer rate when compared to magnetite and manganese-zinc ferrite. Furthermore, the stability analysis was carried out in this investigation to determine which of the two dual solutions is stable. Our thorough results demonstrate that in fluid flow, only the first solution is stable and acceptable.

5.6 ACKNOWLEDGEMENT

The authors sincerely express their gratitude to Universiti Sains Malaysia for the support via the Short-Term Grant (304/PMATHS/6315383).

REFERENCES

Allaw, D., Bachok, N., Md Arifin, N., & Md Ali, F. (2023). Double solutions of unsteady stagnation-point of carbon nanotubes across a permeable exponential stretching/shrinking sheet. *Chinese Journal of Physics*, *85*, 534–552.

Anuar, N. S., Bachok, N., Md Arifin, N., & Rosali, H. (2019). Mixed convection flow and heat transfer of carbon nanotubes over an exponentially stretching/shrinking sheet with suction and slip effect. *Journal of Advanced Research in Fluid Mechanics and Thermal Sciences*, *59*(2), 232–242.

Bachok, N., Ishak, A., & Pop, I. (2012). Boundary layer stagnation-point flow and heat transfer over an exponentially stretching/shrinking sheet in a nanofluid. *International Journal of Heat and Mass Transfer*, *55*(25–26), 8122–8128.

Bahiraei, M., & Hangi, M. (2013). Investigating the efficacy of magnetic nanofluid as a coolant in double-pipe heat exchanger in the presence of magnetic field. *Energy Conversion and Management*, *76*, 1125–1133.

Bezaatpour, M., & Rostamzadeh, H. (2020). Heat transfer enhancement of a fin-and-tube compact heat exchanger by employing magnetite ferrofluid flow and an external magnetic effect. *Applied Thermal Engineering*, *164*, 114462.

Bhattacharyya, K. (2011). Boundary layer flow and heat transfer over an exponentially shrinking sheet. *Chinese Physics Letters*, *28*(7), 074701.

Ebaid, M. S. Y., Ghrair, A. M., & Al-Busoul, M. (2022). Investigation of heat transfer enhancement using ferro-nanofluids (Fe_3O_4 /water) in a heated pipe under the application of magnetic field. *Advances in Mechanical Engineering*, *14*(6). https://doi.org/10.1177/16878132221102647

Eid, M. R., & Nafe, M. A. (2020). Thermal conductivity variation and heat generation effects on magneto-hybrid nanofluid flow in a porous medium with slip condition. *Waves Random Complex Media*, 1–25.

Ghasemian, M., Ashrafi, Z. N., Goharkhah, M., & Ashjaee, M. (2015). Heat transfer characteristics of Fe_3O_4 ferrofluid flowing in a mini channel under constant and alternating magnetic fields. *Journal of Magnetism and Magnetic Materials*, *381*, 158–167.

Ghofrani, A., Dibaei, M., Sima, A. H., & Shafii, M. (2013). Experimental investigation on laminar forced convection heat transfer of ferrofluids under an alternating magnetic field. *Experimental Thermal and Fluid Science*, *49*, 193–200.

Goharkhah, M., Ashjaee, M., & Shahabadi, M. (2016). Experimental investigation on convective heat transfer and hydrodynamic characteristics of magnetite nanofluid under the influence of an alternating magnetic field. *International Journal of Thermal Sciences*, *99*, 113–124.

Hafidzuddin, E. H., Nazar, R., Arifin, N. M., & Pop, I. (2016). Boundary layer flow and heat transfer over a permeable exponentially stretching/shrinking sheet with generalized slip velocity. *Journal of Applied Fluid Mechanics*, *9*(4), 2025–2036.

Jusoh, R., Nazar, R., & Pop, I. (2019). Impact of heat generation/absorption on the unsteady magnetohydrodynamic stagnation point flow and heat transfer of nanofluids. *International Journal of Numerical Methods for Heat and Fluid Flow*, *30*, 557–574.

Lajvardi, M., Moghimi-Rad, J., Hadi, I., Gavili, A., Isfahani, T. D., Zabihi, F., & Sabbaghzadeh, J. (2010). Experimental investigation for enhanced ferrofluid heat transfer under magnetic field effect. *Journal of Magnetism and Magnetic Materials, 322*, 3508–3513.

Magyari, E., & Keller, B. (1999). Heat and mass transfer in the boundary layers on an exponentially stretching continuous surface. *Journal of Physics D: Applied Physics, 32*(5), 577.

Mahmoud, M. A. A., & Waheed, S. E. (2012). MHD flow and heat transfer of a micropolar fluid over a stretching surface with heat generation (absorption) and slip velocity. *Journal of the Egyptian Mathematical Society, 20*, 20–27.

Mat Yasin, S. H., Mohamed, M. K. A., Ismail, Z., & Salleh, M. Z. (2022). Magnetite water based ferrofluid flow and convection heat transfer on a vertical flat plate: Mathematical and statistical modelling. *Case Studies in Thermal Engineering, 40*, 102516.

Mehrez, Z. (2024). Heat transfer enhancement with ferrofluids. In *Advanced materials-based fluids for thermal systems* (pp. 61–100). Elsevier.

Mohd Azmi, H., Mohamed Isa, S. S. P., & Md. Arifin, N. (2020). The boundary layer flow, heat and mass transfer beyond an exponentially stretching/shrinking inclined sheet. *CFD Letters, 12*(8), 98–107.

Najib, N., Bachok, N., Dzulkifli, N. F., & Pop, I. (2022). Numerical results on slip effect over an exponentially stretching/shrinking cylinder. *Mathematics, 10*, 1114.

Nandy, S. K., & Mahapatra, T. R. (2013). Effects of slip and heat generation/absorption on MHD stagnation flow of nanofluid past a stretching/shrinking surface with convective boundary conditions. *International Journal of Heat and Mass Transfer, 64*, 1091–1100.

Papell, S. S. (1965). *Low viscosity magnetic fluid obtained by the colloidal suspension of magnetic particles Patent (Low density and low viscosity magnetic propellant for use under zero gravity conditions).* Patent NumberUS-PATENT-3, 215, 572; US-PATENT-APPL-SN-315096; US-PATENT-CLASS-149-2.

Raj, K., & Chorney, A. F. (1998). Ferrofluid technology—An overview. *Indian Journal of Engineering and Materials Sciences, 5*, 372–389.

Ramli, N., & Ahmad, S. (2021). Stability analysis of magnetohydrodynamic mixed convection flow and heat transfer over a moving flat plate in ferrofluids with suction and slip effects. *Modelling, Simulation and Applications of Complex Systems, CoSMoS 2019, Springer Proceedings in Mathematics and Statistics, 359.* https://doi.org/10.1007/978-981-16-2629-6_12

Rasli, N., & Ramli, N. (2021). Magnetohydrodynamic flow and heat transfer over an exponentially stretching/shrinking sheet in ferrofluids. *Pertanika Journal of Science and Technology, 29*(3), 1811–1829.

Reddy, Y. D., Goud, B. S., Nisar, K. S., Alshahrani, B., Mahmoud, M., & Park, C. (2023). Heat absorption/generation effect on MHD heat transfer fluid flow along a stretching cylinder with a porous medium. *Alexandria Engineering Journal, 64*, 659–666.

Rohni. A., Ahmad, S., & Pop, I. (2012). Flow and heat transfer over an unsteady shrinking sheet with suction in nanofluids. *International Journal of Heat and Mass Transfer, 55*(7–8), 1888–1895.

Roşca, A. V., Roşca, N. C., & Pop, I. (2016). Numerical simulation of the stagnation point flow past a permeable stretching/shrinking sheet with convective boundary condition and heat generation. *International Journal of Numerical Methods for Heat and Fluid Flow, 26*(1), 348–364.

Sheikholeslami, M., & Rashidi, M. (2015). Effect of space dependent magnetic field on free convection of Fe_3O_4-water nanofluid. *Journal of the Taiwan Institute of Chemical Engineers, 56*, 6–15.

Sun, C., Geng, Y., Glowacz, A., Sulowicz, M., Ma, Z., Siarry, P., Gupta, M. K., & Li, Z. (2023). The heat transfer and fluid flow characteristics of water/ Fe_3O_4 ferrofluid flow in a tube with V-cut twisted tape turbulator under the magnetic field effect. *Journal of Magnetism and Magnetic Materials, 586*, 171128.

Wahid, N. S., Md Arifin, N., Khashi'ie, N. S., & Pop, I. (2021). Hybrid nanofluid slip flow over an exponentially stretching/shrinking permeable sheet with heat generation. *Mathematics, 9*, 30. https://doi.org/10.3390/math9010030

Weidman, P. D., Kubitschek, D. G., & Davis, A. M. J. (2006). The effect of transpiration on self-similar boundary layer flow over moving surfaces. *International Journal of Engineering Science, 44*, 730–737.

Yashkun, U., Zaimi, K., Ishak, A., Pop, I., & Sidaoui, R. (2021). Hybrid nanofluid flow through an exponentially stretching/shrinking sheet with mixed convection and Joule heating. *International Journal of Numerical Methods for Heat and Fluid Flow, 31*(6), 1930–1950.

Zainal, N. A., Waini, I., Khashi'ie, N. S., Mohd Kasim, A. R., Naganthran, K., Nazar, R., & Pop, I. (2023). Stagnation point hybrid nanofluid flow past a stretching/shrinking sheet driven by Arrhenius kinetics and radiation effect. *Alexandria Engineering Journal, 68*, 29–38.

CHAPTER 6

Computational Fluid Dynamics Modelling on the Flow Behavior of a High-Speed Train under Crosswinds

Mohammad Arafat and Izuan Amin Ishak

6.1 INTRODUCTION

High-speed trains have transformed modern transportation by providing a fast, efficient and environmentally friendly mode of transit. On the other hand, high-speed trains continue to push the limits of engineering and technology, and understanding their behavior in difficult environmental settings remains a vital problem. The presence of crosswinds is one of the most hazardous conditions affecting high-speed train operation.

Crosswinds, defined as lateral airflows perpendicular to the direction of train velocity, can cause a variety of aerodynamic consequences, possibly risking passenger safety and reducing overall train efficiency. Increased energy consumption, decreased stability, lateral forces and even derailment risks are examples of these impacts. Addressing these issues

DOI: 10.1201/9781032678573-6

necessitates a thorough understanding of the underlying fluid dynamics and the construction of detailed mathematical models.

There are numerous ways available for understanding the aerodynamic performance of a high-speed train, including full-scale experiments, computational approaches using computers, and physical modelling. Each strategy has its pros and cons. The computational fluid dynamics (CFD) technique, in particular, has grown in favor in recent years due to its capacity to compute the entire fluid flow around the train, velocities and pressures. It is expected that CFD capabilities at a reliable level will be adequate to obviate the necessity for physical model tests in many areas.

6.2 HIGH-SPEED TRAIN AERODYNAMICS UNDER CROSSWINDS

In many countries around the world, high-speed trains (HSTs) are regarded as a quick, safe and comfortable mode of transportation. However, there are a number of issues that may cause concern in the coming years including reduction in carbon emissions, extreme weather events, climate change and the demand for high-speed and lightweight structures for vehicles (Baker et al., 2019). The development of new materials introduced lightweight structural designs for different railway applications. The lightweight structure of the next-generation high-speed train (NG-HST) reduces axle loads, which saves money by lowering rail track maintenance costs and the amount of energy needed to drive vehicles. On the other hand, the issue of sudden extreme weather events is becoming unpredictable due to climate change. Climate change-induced increases in the intensity and frequency of extreme weather events can have a significant influence on railway networks. As a result, increasing speed to reduce travel time and the lightweight structural design of the NG-HST raise concerns about the effects of crosswinds on train aerodynamics (Gallagher et al., 2018). Although accidents due to strong crosswinds are not common, they show the possibility and relevance of knowledge of the unsteady flow structures and the resulting acting forces on the surface of the trains.

Many complex flow features are observed in HSTs, including reverse-flow regions, cavities and stagnation points in the gangway and bogie region, laminar-turbulent transition in the nose region, turbulent boundary layer development, separation and re-attachment, and Kàrmàn-vortex streets at the pantographs contact strip (Orellano & Schober, 2006). Nonetheless, only a few basic flow characteristics have a significant impact on crosswind stability.

Figure 6.1 shows a typical flow structure around the train in the crosswind. Generally, when the wind loads affect the windward side of the train, the flow passes towards the leeward side from the roof and underbody area. Vortices roll up to form large vortices in counter-streamwise manners. Later, that dominates the leeward side flow area, with vortices separating on alternate sides along the train body's length.

Crosswinds are caused by differing magnitudes of wind velocities acting in a relative angular direction from the moving train's side. With greater crosswind velocity and lighter material in the NG-HST, the effects of crosswinds on train aerodynamics cannot be avoided. Crosswinds, in principle, could disrupt the vehicle's running stability since aerodynamic forces and moments are magnified. Figure 6.2 illustrates component velocity vectors in relation to a moving train subjected to crosswinds (U_w). β is the natural wind direction determined by the train axis, whereas Ψ is the wind direction relative to the train (i.e. the yaw angle), which is a parameter used to address the various crosswind conditions.

Wind criteria, such as natural wind velocity and wind direction, have been determined to be significantly important in evaluating the safety

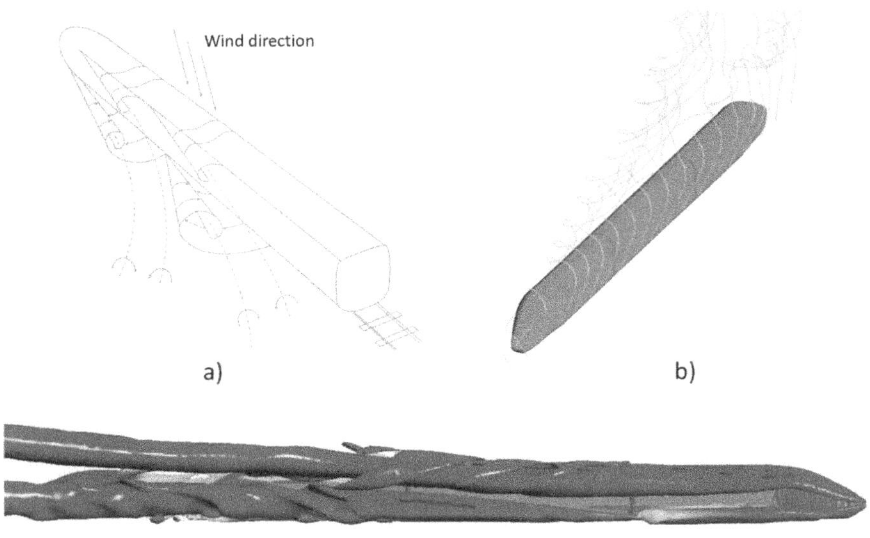

FIGURE 6.1 Vortex at the leeward side of the train in crosswind: (a) flow field (adapted from Copley (1987); (b) streamlines at the leeward side; (c) vortex structure.

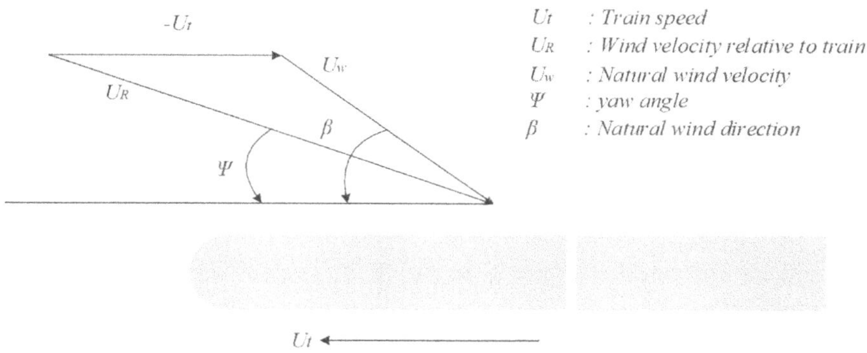

FIGURE 6.2 Natural wind relative to a moving train.

standards of a moving train (Imai et al., 2002). In actuality, the wind angle operating on the vehicle in a crosswind is not necessarily constant as it impacts the side of the railway at a yaw angle (Suzuki et al., 2003). This produces forces (aerodynamic loads) that act on the moving train. Crosswind aerodynamic forces provide a crucial wind load that, in extreme situations, can overturn a vehicle. Furthermore, crosswinds can have an impact on the train's running stability, reducing passenger comfort (Cui et al., 2014).

In this chapter, a CFD methodology was used to explore the flow structure around an HST under crosswinds of 15°, 30°, 45° and 60°. This chapter begins with an introduction and then delves into CFD methodology, including geometry, meshing, simulation setup and numerical methods. Later results are described in terms of velocity streamline, 3D iso-surface and pressure characteristics. Finally, a summary is made based on the simulation results.

6.3 CFD METHODOLOGY

CFD simulation is the main focus of this study as the increasing capacity of modern computational resources (i.e., high-performance computers) lead to the potentiality of CFD in the field of vehicle aerodynamics. This enables the creation of precise train aerodynamic models that capture all relevant flow field characteristics, making it easier to comprehend and solve flow-related problems. The CFD method is a crucial option for resolving complicated flow patterns that are difficult to record using experimental methods. Furthermore, understanding the mechanics of fluid flow around a railway vehicle by solving a bunch of mathematical equations is critical to explaining the occurrence of aerodynamic events.

6.3.1 Geometry Model

In the current study, the NG-HST has been chosen for the investigation (Figure 6.3). The train's three distinguishing features are that it is an HST running at a maximum of 400 km/h, with a double-deck configuration and a lightweight design. Nonetheless, this train concept has been proposed because it achieves goals such as low energy consumption and low life-cycle costs per passenger, even with shorter travel times. The full-scale lengths of the end car and middle car are 21 m and 20 m, respectively (Winter, 2012). A simplified model of NG-HST was used for the simulation consisting of a leading car, a middle car and an end car. Based on the previous study (Fragner & Deiterding, 2018), a 1/25th model scale train was employed where the dimensions are lengths of 840 mm for the end car and 800 mm for the middle car, a height of 186 mm and a width of 122 mm.

6.3.2 Computational Domain and Boundary Conditions

The computational domain of the simulation was designed to capture the flow structure at the wake of the train model. As a result, the computational domain had a length of $32H$, a height of $10H$ and width of $16H$. The domain is large enough to meet the requirements of EN14067-6:2010. The global coordinate point is at the end point of the end car. The ground surface was flat, and the train has a 0.0503 m gap between it and the ground based on the previous study (Arafat & Ishak, 2023; Wang et al., 2021). This ensures that there is sufficient space between the train and the boundary conditions to prevent any unwanted interactions between them and the

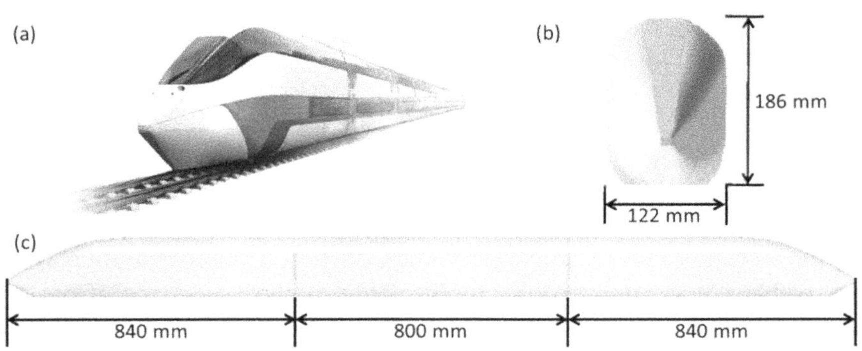

FIGURE 6.3 Next-generation high-speed train developed by German Aerospace Centre (DLR): (a) NG-HST (Winter, 2012); (b) front view; (c) side view.

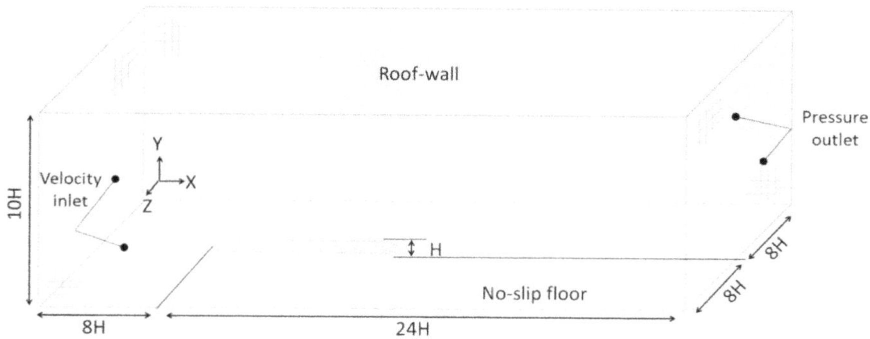

FIGURE 6.4 Boundary conditions of the computational domain used for the crosswind conditions.

flow around the train. The space between the train and the outlet boundary allows the wake pressure to dissipate before reaching the outlet. The pressure outlet is set to zero, denoting gauge pressure as the default operating pressure is 101,325 Pa. Figure 6.4 depicts the specifics of the boundary conditions and enclosure geometry in the simulation.

6.3.3 Solver Setting and Numerical Method

To understand the time-dependent flow physics around the train, the transient mode of the simulation was used in this study. Initial time steps of 0.004 were used in this numerical simulation utilising a shear stress transport (SST)-based delayed detached eddy simulation (DDES) turbulence model (Arafat et al., 2023). The COUPLED approach is applied for pressure-velocity coupling. For spatial discretisation, a second-order upwind approach with a bounded central differencing scheme is used. The entire set of simulation examples was run utilising a parallel computing approach. A summary of the solver simulation setting can be seen in Table 6.1.

TABLE 6.1 Details of Solver Setup Used in the Study

Solver Settings	Description
Mode	Transient
Initial timestep	0.004
Turbulence model	DES (SST)
Pressure velocity-based coupling	COUPLED
Spatial discretisation schemes	Bounded central differencing
Temporal discretisation	Implicit time integration

The governing equation was solved using commercial ANSYS Fluent software. The governing continuity, momentum and energy equations are shown in Eqs. (6.1) to (6.3), respectively.

$$\frac{\partial \rho}{\partial t}+\frac{\partial}{\partial x_j}\left(pu_j\right)=0 \tag{6.1}$$

$$\frac{\partial}{\partial t}\left(pu_i\right)+\frac{\partial}{\partial x_j}\left(pu_iu_j\right)=-\frac{\partial p}{\partial x_i}+\frac{\partial}{\partial x_j}\left[\left(\mu+\mu_t\right)\left(\frac{\partial u_i}{\partial x_j}+\frac{\partial u_j}{\partial x_i}\right)\right] \tag{6.2}$$

$$\frac{\partial\left(\rho E\right)}{\partial t}+\partial\frac{\left(\rho u_jE\right)}{\partial x_j}=\frac{\partial}{\partial x_j}\left(\lambda\frac{\partial T}{\partial x_j}+u_i\tau_{ij}\right)+\rho u_i f_i \tag{6.3}$$

where u_i is the instantaneous velocity component in the i-direction, and t, x_i, p, ρ and μ are the time, spatial coordinates, pressure, the fluid density and the dynamic viscosity, respectively. Moreover, f_i represents any external body forces, and i and j are indices used in tensor notation to indicate different directions. In Eq. (6.3), E is the total energy per unit mass, T is the instantaneous temperature and τ_{ij} is the viscous stress tensor. The turbulent viscosity μ_t is closed by the turbulence models or the DDES equations. The RANS-SST model was developed based on the standard k-ω model to improve the separation prediction in the near-wall region by (Menter, 1994). The equation for turbulent kinetic energy k is

$$\frac{\partial\left(\rho k\right)}{\partial t}+\frac{\partial\left(\rho u_i k\right)}{\partial_{x_i\cdot}}=\frac{\partial}{\partial x_i}\left[\left(\mu+\sigma_k\mu_t\right)\frac{\partial k}{\partial x_i}\right]+Y_k-\beta^*\rho k\omega \tag{6.4}$$

$\beta^*\rho k\omega$ is the dissipation term, representing the rate at which the turbulent kinetic energy is converted into heat through viscous dissipation. Meanwhile, the equation for turbulent dissipation ω is

$$\begin{gathered}\frac{\partial\left(\rho\omega\right)}{\partial t}+\frac{\partial\left(\rho u_i\omega\right)}{\partial x_i}=\frac{\partial}{\partial x_i}\left[\left(\mu+\sigma_\omega\mu_t\right)\frac{\partial\omega}{\partial x_i}\right]+\\ 2\left(1-F_1\right)\rho_{\omega^2}\frac{1}{\omega}\frac{\partial k}{\partial x_i}\frac{\partial\omega}{\partial x_i}+\alpha\rho S^2-\beta\rho\omega^2\end{gathered} \tag{6.5}$$

in which

$$F_1 = \left\{ \left\{ \min\left[\max\left(\frac{\sqrt{k}}{\beta^* \omega y}, \frac{500v}{y^2 \omega} \right), \frac{4\rho\sigma_{\omega^2} k}{CD_{k\omega} y^2} \right] \right\}^4 \right\}, \tag{6.6}$$

$$CD_{k\omega} = \max\left(2\rho\sigma_{\omega^2} \frac{1}{\omega} \frac{\partial k}{\partial x_i} \frac{\partial \omega}{\partial x_i}, 10 \right)$$

$$\mu_t = \frac{\alpha_1 k}{\max(\alpha_1 \omega, SF_2)}, F_2 = \tanh\left\{ \left[\max\left(\frac{\sqrt{k}}{\beta^* \omega y}, \frac{500v}{y^2 \omega} \right) \right]^2 \right\} \tag{6.7}$$

where S is the invariant measure of the strain rate. The constants are calculated by a blend from the corresponding constants via $\alpha = \alpha_1 F_1 + \alpha_2 (1 - F_1)$. Constants are $\beta^* = 0.09$, $\alpha_1 = 5/9$, $\beta_1 = 3/40$, $\sigma_{k1} = 0.85$, $\alpha_2 = 0.44$, $\beta_2 = 0.0828$, $\sigma_{k2} = 1$ and $\sigma_{\omega 2} = 0.856$. In the DES model based on the SST, the dissipation term of the turbulent kinetic energy Y_k in the SST turbulence model (Menter et al., 2003) is modified as

$$Y_k = \rho\beta^* k\omega F_{DES} \tag{6.8}$$

$$F_{DES} = \max\left(\frac{L_t}{C'_{DES} \Delta_{max}}, 1 \right) \tag{6.9}$$

$$L_t = \frac{\sqrt{k}}{\beta^* \omega} \tag{6.10}$$

where the empirical constant C'_{DES} has a value of 0.61. Similarly, the delayed option can also be used to ensure the proper application throughout the boundary layer. The F_{DES} in Eq. (6.9) is modified as

$$F_{DES} = max\left(\frac{L_t}{C'_{DES} \Delta_{max}} (1 - F_{SST}), 1 \right) \tag{6.11}$$

F_{SST} can be selected from 0, F_1 or F_2. Herein, F_1 and F_2 are the blending functions in the SST. The chosen of F_{SST} is introduced in detail by Menter et al. (2003).

6.3.4 Mesh Generation Strategy

The ANSYS Advanced Meshing module was used to generate the hexa-dominant mesh. Details of the mesh are displayed in Figure 6.5. Two refinement boxes were employed to improve the mesh and capture flow structure around and behind the train. The dimensions (L × H × W) of the

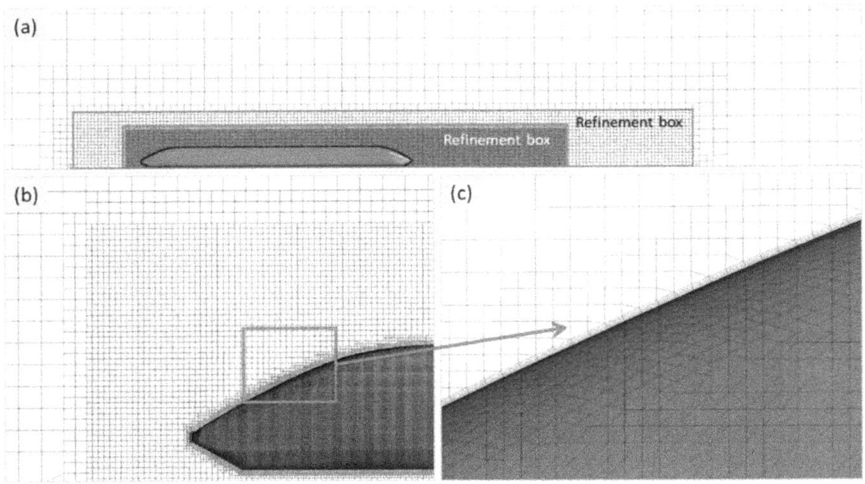

FIGURE 6.5 Mesh resolution used in the case simulation: (a) train model and refinement box; (b) nose of the train; (c) inflation layers on the train surface.

boxes are 8000 × 450 × 800 mm and 3500 × 250 × 300 mm. In addition, on the train's surface, a 5-layer boundary layer mesh is constructed. The first layer thickness was 0.01 mm to keep the y-plus on the train surface within 30 (Medina, 2017). The mesh quality was measured by the two different meshing parameters: skewness was less than 0.38 and the minimum orthogonal quality was 0.208.

The aerodynamic drag coefficient was considered a parameter for the grid convergence study. The drag value difference is smaller between medium and fine mesh compared with coarse and medium mesh. This means that decreasing the mesh element size improves the prediction accuracy. The relative percentage difference between fine and medium is about 0.5%, while for coarse and medium it is around 2%. This means a further improvement in mesh quality will not have much effect on the drag coefficient. Furthermore, even though the medium and fine meshes are adequate to predict aerodynamic loads, an extra fine mesh is required to resolve the flow structures more accurately. As a result, the mesh number in the case simulation was increased to approximately 5.5 million.

To validate the current simulation, numerical results are compared with wind tunnel data from the previous study. In the previous study (Fragner et al., 2015), they employed the RANS model in OpenFOAM and DLR TAU code. The aerodynamic coefficients in the current study are better compared with the CFD simulation results in the previous study. In addition, the relative percentage error of the drag coefficient between experimental and current simulation work is about 7%, which is reasonable.

6.4 RESULTS

6.4.1 Velocity Streamlines

Velocity streamlines are used to visualise fluid flow motions in the domain that are tangential to the instantaneous velocity direction. In this section, velocity streamlines with pressure contours of crosswind conditions are discussed. Figure 6.6 depicts velocity streamlines for the yaw angles of 15°, 30°, 45° and 60°. Four cross-section planes were drawn to visualise the streamline with pressure in four different locations along with the x-axis of the computational domain.

For yaw angles of 15° and 30°, two major vortices are formed on the leeward side, as shown in Figures 6.6a and 6.6b. The center of the vortices

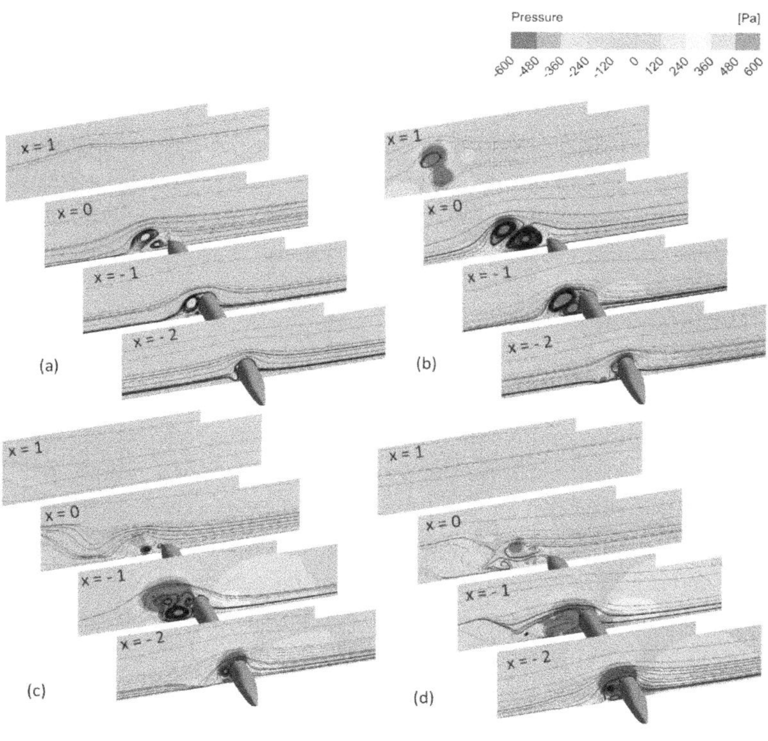

FIGURE 6.6 Streamlines with pressure contour at different yaw angles: (a) 15°; (b) 30°; (c) 45°; and (d) 60°.

shows low pressure, which leads to the leeward side being a dominant negative pressure region. In contrast, the windward side shows high pressure due to the train surface, making it a positive pressure-dominated area. Due to the pressure difference, the vortices are mainly initiated at the head car. Additionally, fluid motion on the bottom and top edges of the leeward side develops two major counter-rotating vortices. The circulation increases as it goes further from the nose in a backward direction. It can also be seen that the vortices separate from the train surface when it is near the end of the train. In addition, it is noticeable that the vortices are larger at the 30° yaw angle compared to 15°.

On the other hand, velocity streamlines for 45° and 60° yaw angles are completely different from the lower yaw angles (15° and 30°). This can be seen in Figures 6.6c and 6.6d. It develops complex vortices on the leeward side and at the wake of the train. This indicates that with increasing yaw angles, the flow field on the leeward side of the train will become more complex and unsteady, which may have an impact on the trackside infrastructure (Ishak et al., 2019). Besides, Figure 6.6c shows that the pressure area increased significantly as the wind angle reached 45°. In addition to the two major vortices, some other small vortices also formed due to the train boundary layer. As a result of increasing the wind angle to 60°, the negative pressure region is increased on the leeward side, and positive pressure on the windward side can be seen in Figure 6.6d. This suggests the complexity of the flow structure at the leeward side of the NG-HST and indicates a possible impact around the train with an increasing crosswind angle.

6.4.2 Instantaneous Vortex Core

Figures 6.7 and 6.8 represent the iso-surface of crosswind conditions at four yaw angles ranging from 15° to 60° at time $t = 3.59$s. Figure 6.7a shows that a total of three major vortices are formed on the leeward side of the train due to the wind loads on the train surface. Vortex V_1 and V_2 originate from the bottom and top edges of the head car, respectively, while V_3 originates from the edge and inter-car gap between the head and middle car.

A counter-rotating vortex core can be seen at a yaw angle of 30°. It is formed in the underbody area of the nose and develops further as the train moves forward. When wind loads approach the train's windward side, the flow from the roof and underbody area flows to the leeward side. Then the vortices roll up in a counter-streamwise direction to form massive vortices. Later, vortices separate on alternate sides of the train, dominating the leeward side flow area.

For the yaw angles of 45° and 60°, the vortex formation is more unsteady and complex. At this point, a large number of different-sized vortices are formed on the leeward side. The height of the vortex structure is almost the same; however, width and density are relatively large in 60° crosswind conditions. It can be seen that the vortex formation is mainly generated in the underbody and roof areas of the train. With the increase in yaw angles, there are several changes in the vortex structure around NG-HST that are mentionable. As the yaw angle increases, the vortex length at the train's wake decreases, while the vortex width on the train's leeward side increases.

Figure 6.8 shows the vortex formation around the train at different wind angles from the top. This figure shows the zone where counter-rotating vortices are formed, including the vortex angle at the wake with respect to the train (Table 6.2). At a 30° yaw angle, the vortex started to form a counter-rotating like shape and the formation started at the inter-car gap between the head car and the middle car shown in Figure 6.8b. However, for a 45° wind angle the similar formation initiated earlier means the middle part of the head car.

Based on the data found from four different yaw angles and vortex angles at the wake, it is possible to formulate an equation. However, it is required to follow the same simulation conditions as used in the current study. The vortex angle at the wake can be predicted using Eq. (6.12), which is expressed as:

$$Y = f(x) \approx 0.195241\Psi^{1.34553} - 1.34474 \tag{6.12}$$

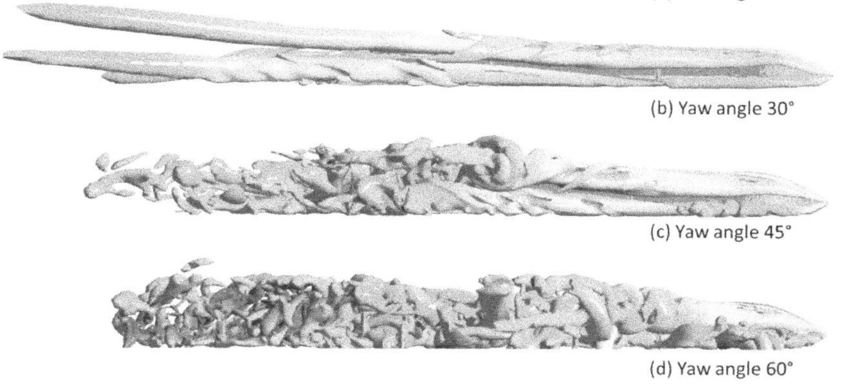

FIGURE 6.7 Vortex core for steady crosswind at different yaw angles at $t = 3.59$s (side view).

TABLE 6.2 Vortex Angle at the Wake with Different Steady Crosswind

Steady Wind Angle	Vortex Angle at the Wake
15°	6°
30°	18°
45°	31°
60°	47°

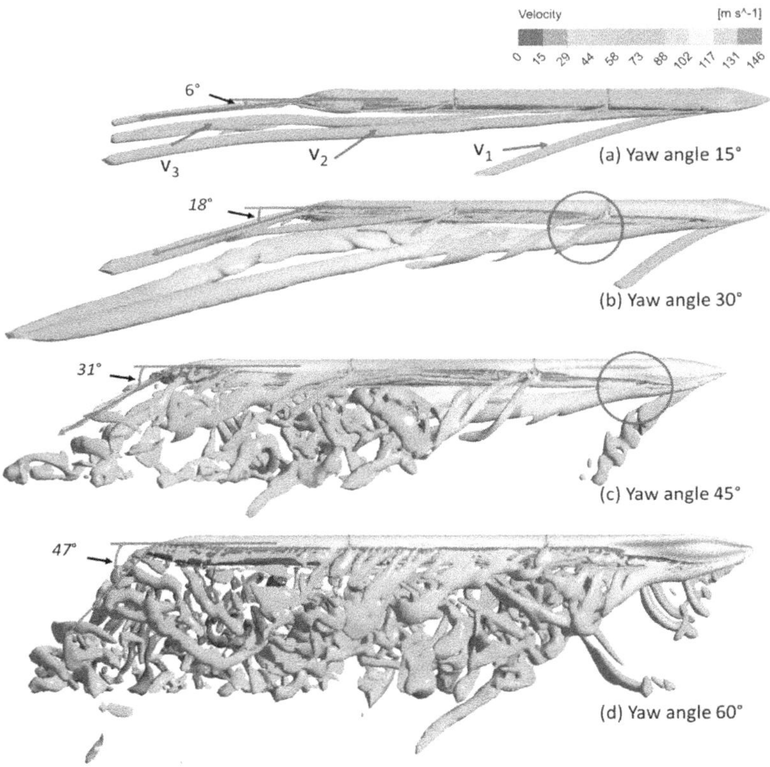

FIGURE 6.8 Vortex formation for steady crosswind at different yaw angles t = 3.59s (top view).

where Ψ is the wind angle at the train surface. The power curve fitting function was used to formulate the equation.

Figure 6.8d shows the vortex formation at the leeward side of the train for a wind angle of 60°. Vortices on the leeward side are more complex and unsteady. In addition, vortices are getting smaller in shape and are formed from the roof and underbody area of the leeward side due to the wind loads on the windward side of the NG-HST. Even though these results provide a significant contribution to vortex structure formation around the NG-HST, it is suggested that further study is required to understand other properties that have an impact on vortex structure formation.

6.4.3 Pressure Characteristics

Pressure distribution is one of the important characteristics in train aerodynamics-related research. The difference in pressure distribution caused by winds at different yaw degrees may result in a noticeable variance in the aerodynamic loads on the train, particularly for lift and side forces. In this section, pressure distribution in crosswinds at different wind angles is revealed using pressure contours.

Figure 6.9 shows the pressure contours for the crosswind conditions (15° and 30°) on the NG-HST. A crosswind angle of 15° depicts the common phenomenon of pressure distribution, as described earlier. Positive pressure dominates at the train's nose, with negative pressure on the train's leeward side. In addition, negative pressure spreads in some areas on the leeward side of the train's wake.

At a 30° yaw angle, positive pressure is spread almost equal to the train length on the windward side. Positive pressure can also be seen at the wake and leeward wake. On the other hand, the leeward side and near-far wake of the train are dominated by negative pressure. It also shows that the longest length negative, along with the train length, can be seen at a crosswind angle of 30°.

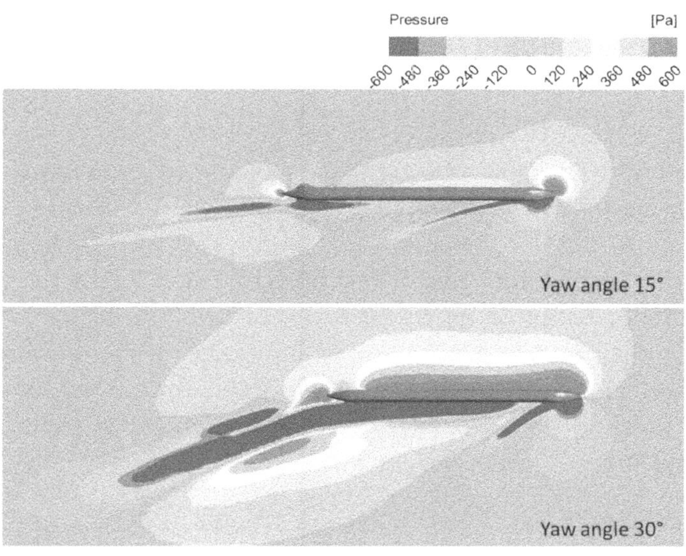

FIGURE 6.9 Pressure contour at lower yaw angle conditions for steady crosswind at $y = -0.25H$.

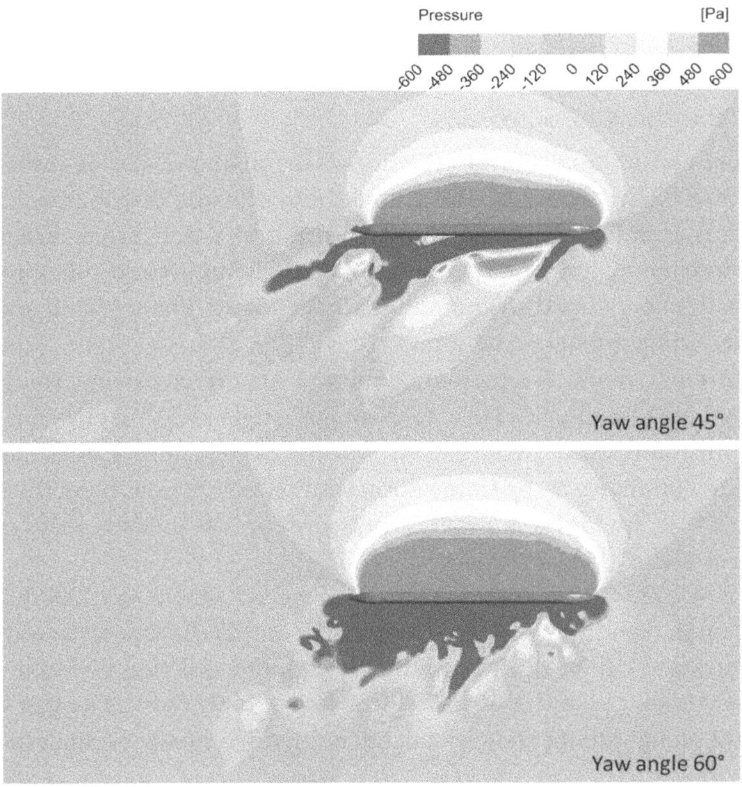

FIGURE 6.10 Pressure contour at higher yaw angle conditions for steady crosswind at $y = -0.25H$.

For relatively higher crosswind angles (45° and 60°) pressure contour is shown in Figure 6.10. It can be seen that the windward side is completely dominated by positive pressure, while negative pressure is dominant on the leeward side of the train. For a 60° crosswind angle, the pressure distribution situation is nearly identical. In addition, as the yaw angle increases, the pressure distribution is dominated by the windward and leeward sides for positive and negative pressure, respectively. The pressure contour proves the complexity of vortex formation due to the increase in crosswind angle shown in the previous section.

6.5 SUMMARY

In this chapter, four crosswind angles were investigated in order to understand the flow structure around a high-speed train under crosswind conditions using a CFD approach.

The results show that the velocity streamlines for 45° and 60° yaw angles are completely different from the lower yaw angles (15° and 30°). It forms complicated vortices on the leeward side and wake of the train. This means that as the yaw angle increases, the flow field on the train's leeward side becomes more complex and unsteady. Furthermore, when yaw angles increase, significant changes in the vortex formation around NG-HST are visible. The changes are that the vortex length at the wake decreases as the yaw angle increases and the vortex width on the train's leeward side increases. Additionally, at a 30° yaw angle, the vortex started to form a counter-rotating shape and vortex formations were getting more complex and unsteady with the increase in yaw angle. Furthermore, pressure distribution showed the low-pressure region moving towards the leeward side with the increase in crosswind angle.

6.6 ACKNOWLEDGEMENT

This research was financially supported by the Ministry of Higher Education (MOHE), Malaysia through Fundamental Research Grant Scheme (FRGS/1/2022/TK02/UTHM/03/1).

REFERENCES

Arafat, M., & Ishak, I. A. (2023). Aerodynamic performance of a next-generation high-speed train under transient crosswind. *OSF Preprint*. https://doi.org/10.31219/osf.io/u3mqb

Arafat, M., Ishak, I. A., Mohammad, A. F., Khalid, A., Mohmad Jaát, M. N., & Yasak, M. F. (2023). Effect of Reynolds number on the wake of a next-generation high-speed train using CFD analysis. *CFD Letters*, *15*(1), 76–87. https://doi.org/10.37934/cfdl.15.1.7687

Baker, C., Johnson, T., Flynn, D., Hemida, H., Quinn, A., Soper, D., & Sterling, M. (2019). Train aerodynamics. In *Train aerodynamics: Fundamentals and applications*. Elsevier. https://doi.org/10.1016/C2016-0-04444-3

Copley, J. M. (1987). The three-dimensional flow around railway trains. *Journal of Wind Engineering and Industrial Aerodynamics*, *26*(1), 21–52. https://doi.org/10.1016/0167-6105(87)90034-1

Cui, T., Zhang, W., & Sun, B. (2014). Investigation of train safety domain in cross wind in respect of attitude change. *Journal of Wind Engineering and Industrial Aerodynamics*, *130*, 75–87. https://doi.org/10.1016/j.jweia.2014.04.006

Fragner, M., & Deiterding, R. (2018). Investigating side-wind stability of high speed trains using high resolution large eddy simulations and hybrid models. In *Computational methods in applied sciences* (Vol. 45, pp. 223–241). https://doi.org/10.1007/978-3-319-54490-8_14

Fragner, M., Weinman, K. A., Deiterding, R., Fey, U., & Wagner, C. (2015). Comparison of industrial and scientific CFD approaches for predicting cross wind stability of the NGT2 model train geometry. *International Journal of Railway Technology*, *4*(1), 1–28. https://doi.org/10.4203/ijrt.4.1.1

Gallagher, M., Morden, J., Baker, C., Soper, D., Quinn, A., Hemida, H., & Sterling, M. (2018). Trains in crosswinds—comparison of full-scale on-train measurements, physical model tests and CFD calculations. *Journal of Wind Engineering and Industrial Aerodynamics*, *175*, 428–444. https://doi.org/10.1016/j.jweia.2018.03.002

Imai, T., Fujii, T., Tanemoto, K., Shimamura, T., Maeda, T., Ishida, H., & Hibino, Y. (2002). New train regulation method based on wind direction and velocity of natural wind against strong winds. *Journal of Wind Engineering and Industrial Aerodynamics*, *90*(12–15), 1601–1610. https://doi.org/10.1016/S0167-6105(02)00273-8

Ishak, I. A., Ali, M. S. M., Sakri, F. M., Zulkifli, F. H., Darlis, N., Mahmudin, R., Abidin, S. F. Z., Azizul, M. A., Sulaiman, S., Samiran, N. A., & Khalid, A. (2019). Aerodynamic characteristics around a generic train moving on different embankments under the influence of crosswind. *Journal of Advanced Research in Fluid Mechanics and Thermal Sciences*, *61*(1), 106–128.

Medina, A. (2017). *High-speed train aerodynamics under crosswinds: From standard wind tunnel testing to the effect of the train cooling systems* [Federal University of Uberlândia]. https://repositorio.ufu.br/bitstream/123456789/20351/3/HighspeedTrainAerodynamics.pdf

Menter, F. R. (1994). Two-equation eddy-viscosity turbulence models for engineering applications. *AIAA Journal*, *32*(8), 1598–1605. https://doi.org/10.2514/3.12149

Menter, F. R., Kuntz, M., & Langtry, R. (2003). Ten years of industrial experience with the SST turbulence model. *Turbulence Heat and Mass Transfer*, *4*, 625–632.

Orellano, A., & Schober, M. (2006). Aerodynamic performance of a typical high-speed train. *4th WSEAS international conference on fluid mechanics and aerodynamics*, *2006*, 18–25. www.wseas.us/e-library/conferences/2006elounda2/papers/538-183.pdf

Suzuki, M., Tanemoto, K., & Maeda, T. (2003). Aerodynamic characteristics of train/vehicles under cross winds. *Journal of Wind Engineering and Industrial Aerodynamics*, *91*(1–2), 209–218. https://doi.org/10.1016/S0167-6105(02)00346-X

Wang, S., Wang, R., Xia, Y., Sun, Z., You, L., & Zhang, J. (2021). Multi-objective aerodynamic optimization of high-speed train heads based on the PDE parametric modeling. *Structural and Multidisciplinary Optimization*, *64*(3), 1285–1304. https://doi.org/10.1007/s00158-021-02916-0

Winter, J. (2012). Novel rail vehicle concepts for a high speed train: The next generation train. *Civil-Comp Proceedings*, *98*. https://doi.org/10.4203/ccp.98.22

CHAPTER 7

Numerical Solution of Linear Simultaneous Equations Based on Particle Swarm Optimisation of Multi-Agent System

Sora Nakao and Hirohide Haga

7.1 INTRODUCTION

In this article, we propose a method for numerical solution of linear simultaneous equations as a new application of multi-agent simulation (MAS; Wilensky & Rand, 2015), which is often used in the field of simulation. In this study, we use particle swarm optimisation (PSO; Hassanien & Emary, 2015), which is an optimisation method based on the concept of MAS to implement the proposed method. In addition, several improvements of execution speed-up were introduced into PSO to deal with higher dimensional (multi-variable) problems.

MAS is a system that accomplishes tasks that are difficult, such as mathematical analysis. It accomplishes tasks by the interaction among

DOI: 10.1201/9781032678573-7

multiple agents. Each agent in MAS has different characteristics, making it suitable for simulating groups with unique characteristics. For this reason, MAS is often used to simulate complex phenomenon such as elections and traffic congestion. Furthermore, there are optimisation algorithms named particle swarm optimisation (PSO) based on the concept of MAS (Czarnowski et al., 2013), and MAS is also used in fields such as data mining and machine learning. Thus, MAS is now used in various fields and is expected to play an active role in new fields in the future.

One mathematical application of MAS is already reported in the field of numerical simulation of differential equations (Shimizu & Haga, 2018). In this chapter, we propose the method to get the numerical solution of linear simultaneous equations as a new application of MAS. Linear simultaneous equations are important equations in engineering, but until now, no method using MAS has been proposed. If MAS can be applied to solve linear simultaneous equations, it will be possible to use MAS in various fields such as kinematics and electric circuit design, further expanding the scope of MAS applications.

In this research, we adopted PSO for numerical computation of linear simultaneous equations. PSO is an optimisation algorithm based on the MAS concept. PSO is a model that mimics a group of swarms of insects searching for food by exchanging information with each other. The model is used for continuous-valued optimisation and combinatorial optimisation problems (Abraham et al., 2006). In PSO, many search agents called *particles* are put in the search space, and while exchanging the evaluation information of positions with each other, they move to the solution by moving to a position with higher evaluation value. The features of PSO are that

1. The algorithm itself is simple and easy to implement.
2. Convergence is fast because the search area is gradually reduced.
3. It is rare to fall into a local solution because it does not use a gradient, as in the hill-climbing method.

Since the amount of computing for numerical calculations of linear simultaneous equations increases significantly as the number of variables increases, PSO was used as the method for this study because of its fast convergence. In this study, the PSO solution method was realised by designing the adaptivity function so that the evaluated value of the particle is the sum of the squares of the errors between the left-hand side values and the right-hand side values when the current position of the particle is substituted into the given linear simultaneous equation. Since it is difficult for ordinary PSO to fully demonstrate its performance in high-dimensional spaces, we rearranged the particles in search space according to the average speed of particles in search space, and instead of using ordinary uniform random numbers for the placement of particles, we used the Cauchy distribution (Feller, 1971), which has been shown to be effective in both intensive search and global search in previous research (Kaji, 2014). In this chapter, we describe a specific implementation of the numerical computation method for linear simultaneous equations based on the concept of MAS using PSO, including these improvements.

In this study, we were able to realise a MAS-based solution method by using PSO. However, even with some speed-up methods, the method can only solve linear systems of up to 40 dimensions in real time, and it is difficult to apply the method to large-scale linear systems of thousands or tens of thousands of variables which are actually used in real life. This is due to be because the number of particles sufficient for the performance of PSO increases exponentially, and thus the number of particles is completely insufficient in higher dimensional spaces. To cope with higher dimensional spaces, it is considered necessary to improve the performance of the PSO by some kind of pre-processing for the given problem, but this is a subject for future research.

7.2 BACKGROUND OF RESEARCH

7.2.1 Overview of MAS

MAS is a system in which multiple agents interact with each other to accomplish tasks that are difficult to solve by individual agents alone and is often used to simulate human society. In MAS, each agent has different characteristics, making it suitable for simulations of groups with unique characteristics. For example, MAS is used to simulate natural phenomena such as forest fires and food chains in addition to human societies (Wilensky & Rand, 2015). Furthermore, in the field of swarm intelligence (Mine et al., 1996), there are optimisation algorithms that model swarm behavior of living organisms, and they are expected to be used for data mining, pattern recognition and document classification.

As described earlier, MAS is used in various fields other than social simulation and is expected to play an active role in new fields in the future. We propose a numerical method for linear simultaneous equations as a new application. Linear simultaneous equations are important equations in engineering, and are often used, for example, in kinematics and electric circuit design. In other words, if MAS can be used for solving linear simultaneous equations, it will be possible to use MAS in various fields, thereby expanding the field of application of MAS. In this study, we examine the solution of linear systems of simultaneous equations by MAS and discuss the advantages and disadvantages of this method.

7.2.2 Overview of Particle Swarm Optimisation (PSO)

PSO is a kind of continuous-valued optimisation algorithm based on MAS, which models the movement of organisms as they search for food in groups (Abraham et al., 2006). The algorithm is characterised by the simplicity of the algorithm itself, which makes it easy to implement, and as it does not use gradients in the search such as in the hill-climbing method (Russel & Novig, 2003), which makes it difficult to fall into local solutions. In addition, since the search is conducted in such a way that the search area gradually narrows, it is faster than genetic algorithms in finding a solution.

In PSO, many agents called *particles* move freely in the multidimensional virtual space of the search target. The particles have *velocity* and *position* parameters, and approach to the final solution by moving through the space while exchanging their current position information with each other. The position of a particle at time $t + 1$ is determined by its velocity $v(t)$ as follows:

$$x(t+1)=x(t)+v(t) \tag{7.1}$$

Note that, of course, both velocity $v(t)$ and location $x(t)$ are n-dimensional vectors. The velocity is determined by the mutually independent random numbers r_1, r_2, the best position coordinate $\hat{p}$ found by the particle, and the best position coordinate $\widehat{p_g}$ found by the whole swarm as in the following equation. Note that w is called the *inertia factor* and is used to balance the global search and local search:

$$v(t+1)=wv(t)+c_1r_1\left(\hat{p}-x(t)\right)+c_2r_2\left(\widehat{p_g}-x(t)\right) \tag{7.2}$$

where the term including $\hat{p}$ is called the *cognitive component* and the term including $\widehat{p_g}$ is called the *sociability component*. The former is the component that adjusts the speed influenced by the optimal position. This

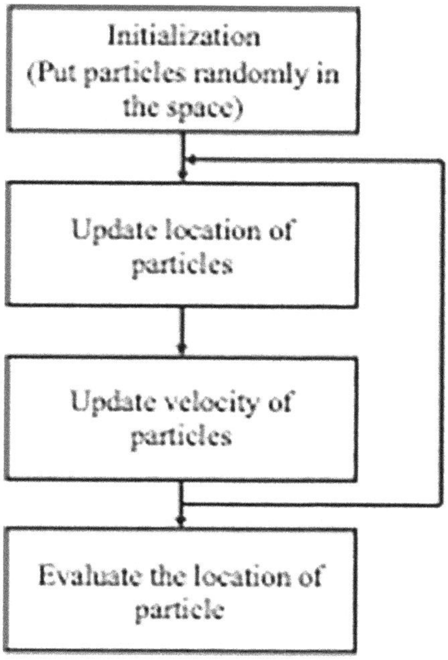

FIGURE 7.1 Flowchart of general PSO processing.

optimal point is found by one particle. The latter is the part that adjusts the speed influenced by the optimal position which is found by a group. In addition, c_1 and c_2 in Eq. (7.2) are coefficients that adjust the balance between the cognitive and social components. Using these equations, PSO is executed. Figure 7.1 shows the flowchart of PSO execution.

First, particles are randomly placed in the space to be searched as an initialisation process, and the initial value of the velocity is set to 0. Next, the position is updated based on the current velocity by using Eq. (7.1), and the velocity is calculated at the new position by using Eq. (7.2). Then, whether the current position is close to the solution is evaluated by an evaluation function called the *adaptivity function*. By repeating this process of updating the position and velocity and evaluating the position, the solution is found.

7.2.3 Some Issues of PSO Method

In PSO, as the algorithm itself does not require mathematical knowledge of the given problem (no substitutions, factorisations or other mathematical operations will be executed), it seems to be easy to apply problems that are

TABLE 7.1 Efficiency and Accuracy of PSO in Higher-Dimensional Space

Dimension	Exact Solution	Approximate Solution
2	0.0	0.00
50	0.0	118.39
100	0.0	22,100.5
200	0.0	115,847.0

mathematically difficult to solve. Furthermore, the search performance is relatively effective when a sufficient number of particles are available in the search space. However, in high-dimensional search spaces where a sufficient number of particles is not available, the search becomes like a brute-force search, and the search efficiency drops significantly, converging to an inferior solution at an early stage because it does not use knowledge. Table 7.1 shows the performance evaluation of PSO in high dimensions shown in previous studies (Parsopoulos & Vrahatis, 2004). The number of particles is set to 500, the maximum number of iterations is 1000 and the approximate solutions of the Rastrigin function (a multimodal function used to evaluate the performance of optimisation algorithms; Hoffmeister & Bäck, 1991) from 2 to 200 dimensions are obtained by PSO.

7.3 ACCELERATING PSO

7.3.1 Using Cauchy Distribution

In the case of ordinary PSO, the number of particles is not sufficient in high-dimensional spaces, and the search efficiency is reduced. In this article, we introduce the Cauchy distribution to improve this problem.

First, we explain particle movement based on the Cauchy distribution. The Cauchy distribution (Feller, 1971) is a continuous probability distribution represented by a probability distribution function as in Eq. (7.3), and has the shape shown in Figure 7.2. It is a bell-shaped distribution like the normal distribution, but it has the characteristic of having a "long tail" or hem that continues indefinitely.

$$p(x;\mu,\gamma)=\frac{1}{\pi}\left[\frac{\gamma}{(x-\mu)^2+\gamma^2}\right] \tag{7.3}$$

In ordinary PSO, particles are placed by using a uniform random function, so they are placed uniformly random in the search range. Therefore, it is necessary to set the entire range where a solution can exist as the search

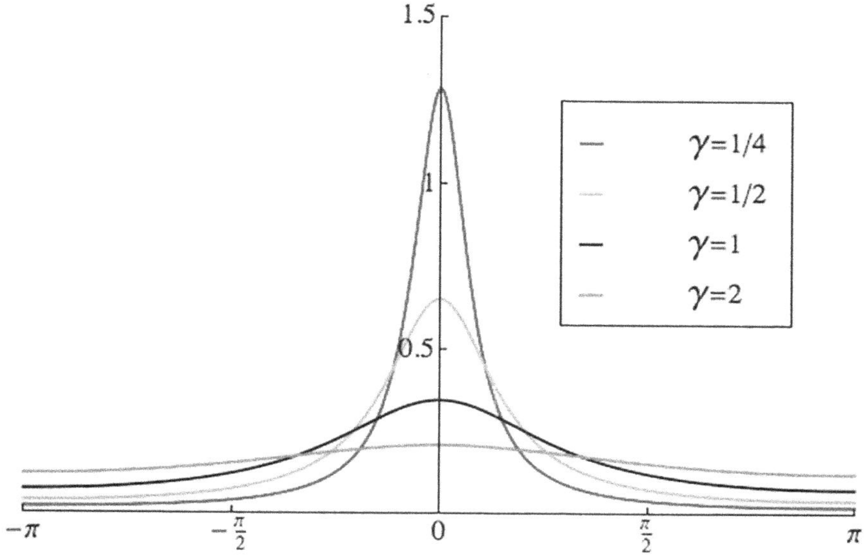

FIGURE 7.2 Cauchy distribution function.

TABLE 7.2 Evaluation of CPSO

Iteration	By PSO	By CPSO
500	161,116	10674
1000	122,511	6334
2000	113,569	3684
3000	111,733	2844
4000	111,164	1987
50000	110,956	1505

range. However, the Cauchy distribution generates random numbers not only within the set range, but also a small number of random numbers outside the set range, so that particles can be concentrated in the set search range and simultaneously placed outside the range. As a result, both local search and global search can be achieved while minimising particle wastage, and performance in high dimensions can be improved. Note that the term "wastage" means the particles which are not used for optimisation. Table 7.2 shows the performance evaluation of PSO employing the Cauchy distribution (hereinafter CPSO) shown in the previous study. It shows the approximate solution (evaluation value) of the Rastrigin function in 200 dimensions with 500 particles obtained by PSO and CPSO at each number of iterations. The CPSO shows a continuous improvement of the solution.

7.3.2 Relocation of Particles

Next, we discuss the necessity of particle rearrangement. As mentioned before, PSO has the disadvantage. It converges to an inferior solution early on in high-dimensional spaces, and the search cannot proceed further. The simplest solution to avoid this convergence to an inferior solution is to rearrange the particles when it converges to an inferior solution and start the search again. In other words, the search continues until a solution with the desired accuracy is obtained, and if the time cost is ignored, the discovery of a solution can be guaranteed regardless of the number of particles. When computation does not stop, it should be terminated in certain time. In such a case, we will be able to get better values and will start computation again by using such "better" values as initial values. Of course, we will "tackle" the increase of computing time. This problem will be able to solve partly by using hardware advancement such as the usage of parallel computing environment. Figures 7.3 and 7.4 show images of the approach from an inferior solution to a true solution by rearrangement.

Figure 7.3 shows an image of particles converging to the inferior solution. Because the initial placement is biased, the particles do not swap their positions as they approach the best position, and the search does not proceed because the particles only gather at the best position at the initial placement stage. (In reality, such a situation rarely occurs in two-dimensional space. In fact, in two-dimensional space, this situation rarely occurs, but in higher dimensional space, such a situation is often observed.) However, by rearranging the position, as shown in Figure 7.3, a new best position can be found, and the search can proceed.

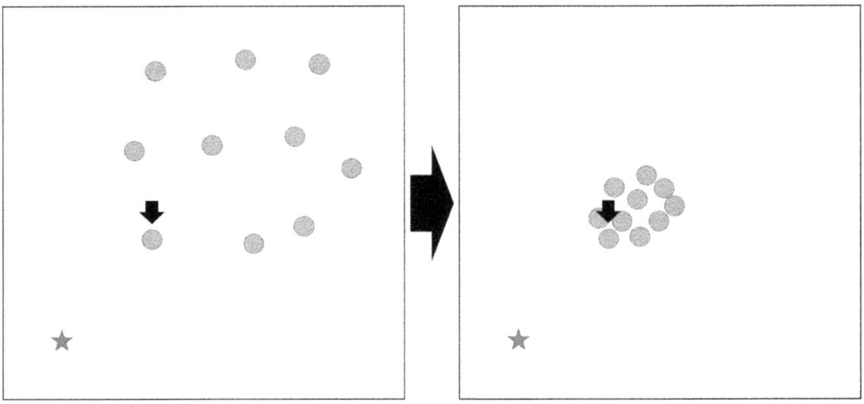

FIGURE 7.3 Convergence to inferior solution.

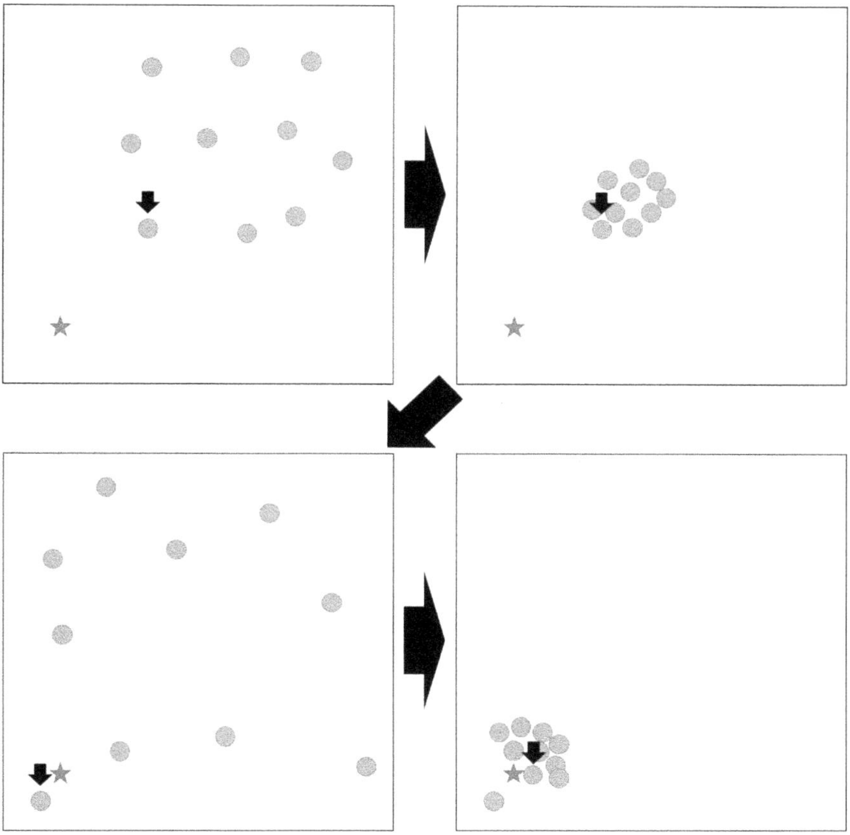

FIGURE 7.4 Effect of rearrangement which avoids converging inferior solution.

In such a method, all particles are reallocated, and the search is executed again. For example, in one previous study, a predator agent was introduced to select particles to be reallocated, and only particles that were preyed upon by the predator were reallocated, thereby enabling both search near inferior solutions and search at distant locations.

7.4 PROPOSED METHOD

In order to implement numerical computation of linear simultaneous equations by MAS, we focused on the swarm intelligence (artificial intelligence technology based on biological swarms), which is closely related to MAS. In the field of swarm intelligence, there are several optimisation algorithms that model swarms of organisms, but in this study, we focused on the convergence speed and the avoidance of convergence to a local solution such as a multimodal function of PSO because the computation time increases significantly with the number of variables in the numerical computation of linear simultaneous equations.

In PSO, the evaluation of the agent's position is performed using an *adaptive function*, and since we are dealing with linear simultaneous equations (Eq. (7.5)), we designed the adaptive function so that the evaluated value is the sum of the squared errors of the left and right sides, as in Eq. (7.4).

$$score = \sum_{i=0}^{n} \left(f_i(x) - b_i \right)^2 \tag{7.4}$$

$$\begin{aligned} f_0(x) &= a_{00}x_1 + a_{01}x_2 + \cdots + a_{0m}x_m = b_0 \\ f_1(x) &= a_{10}x_1 + a_{11}x_2 + \cdots + a_{1m}x_m = b_1 \\ &\cdots \\ f_n(x) &= a_{n0}x_1 + a_{n1}x_2 + \cdots + a_{nm}x_m = b_n \end{aligned} \tag{7.5}$$

Furthermore, to improve the performance in high-dimensional spaces, the following two improvements were applied to the regular PSO.

- Particle placement is based on the Cauchy distribution, enabling both intensive and global searches while minimising particle wastage.
- Rearrangement of particles is introduced to avoid stopping the search due to convergence to inferior solutions.

In addition, since the linear simultaneous equations have a unique solution, it is assumed that the true solution is located around the current best position (inferior solution), so that the relocation is executed around the current best position. Furthermore, the relocation interval is approaching the average of particle velocities in the swarm instead of a fixed interval. As a result, relocation is automatically performed when the whole swarm becomes slow, thus reducing the wasteful use of relocation intervals. Figure 7.5 shows an overview of the PSO procedure proposed in this study (for the sake of simplicity, the termination decision is omitted). First, particle placement based on the Cauchy distribution is performed, followed by updating of particle velocities and positions, evaluation of positions and determination of swarm halt (based on the average velocity of particles in the county), and the process is repeated until a solution with a predetermined accuracy is obtained.

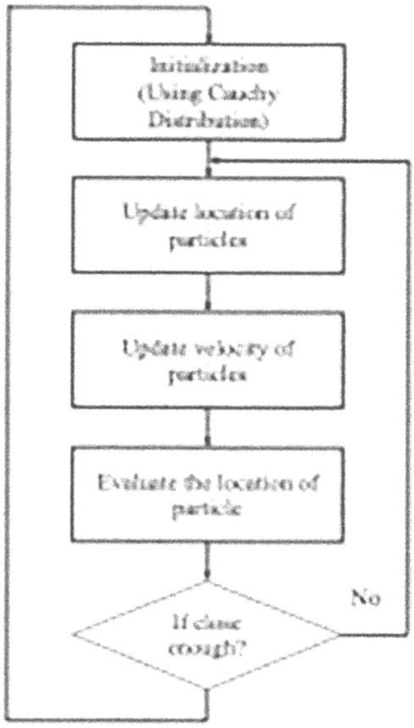

FIGURE 7.5 Procedure of PSO used in this research.

7.5 IMPLEMENTATION

7.5.1 Software Structure

In this study, PSO is designed in an object-oriented manner. Figure 7.6 shows the class diagram used in our implementation. Methods and values are shown in this diagram and are not exhaustive (e.g., methods for displaying results). The green-colored parts mean the class variables and the white-colored parts indicate the methods.

Here, Sim is the main class for executing PSO, and Swarm and Particle are classes for groups and particles, respectively. Param is the class of parameters such as inertia factor and acceleration coefficient. Although there is no problem in using common fixed values for the parameters, we designed the Param class in consideration of future extensions such as having different parameters for each particle and dynamic control of the parameters. Similarly, we designed the Sim class as a superclass of Swarm to enable future extensions such as search by swarms with two different characteristics and made it possible to control Swarm from the Sim class. The evaluation value update method, which is the particle's method in the figure, is the adaptivity function, and is set as the sum of the squares of the errors between the left-hand side value and the right-hand side value when

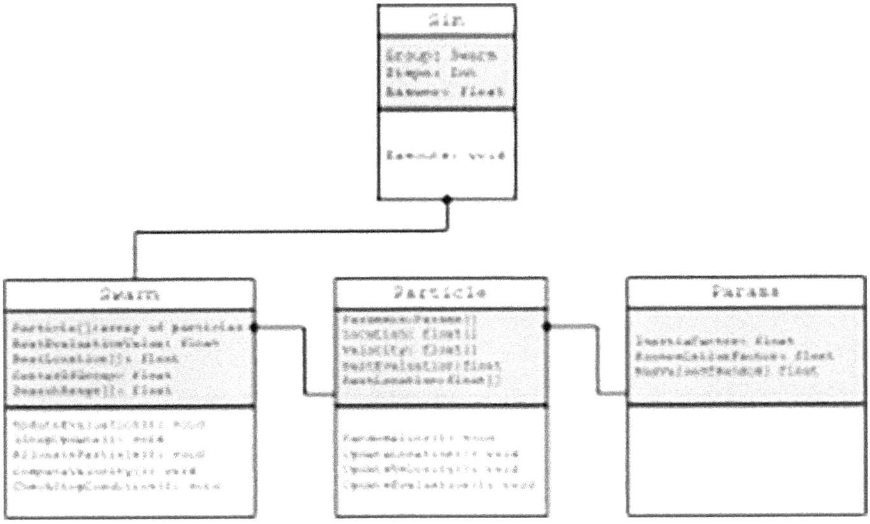

FIGURE 7.6 Class diagram of implementation.

the coordinates of the particle's current position are substituted into the given linear simultaneous equation as described earlier.

7.5.2 Implementation of Particle Arrangement Based on Cauchy Distribution

In a previous study (Simon Fraser University, 2013), it was shown that the performance in high-dimensional spaces can be improved by moving particles using the Cauchy distribution. In this program, we changed the placement of particles from uniform random numbers to the Cauchy distribution-based placement. By using the Cauchy distribution for the placement of particles, the true solution can be found even when the true solution does not exist in the pre-defined search range, and when it does exist, convergence can be accelerated by concentrated placement of particles. Figure 7.7 shows the effect of placement using the Cauchy distribution. Note that the blue star is a solution, the black arrow is an optimal location and the orange circles represent particles. The left panel of the figure shows placement by uniform random numbers, which converges to the best position at the initial point in time at the edge of the set search space because there is no true solution within the range. However, by using the Cauchy distribution, as shown on the right, a small number of particles are placed outside the search range, allowing the true solution to be found. The number of particles outside the search range can be controlled

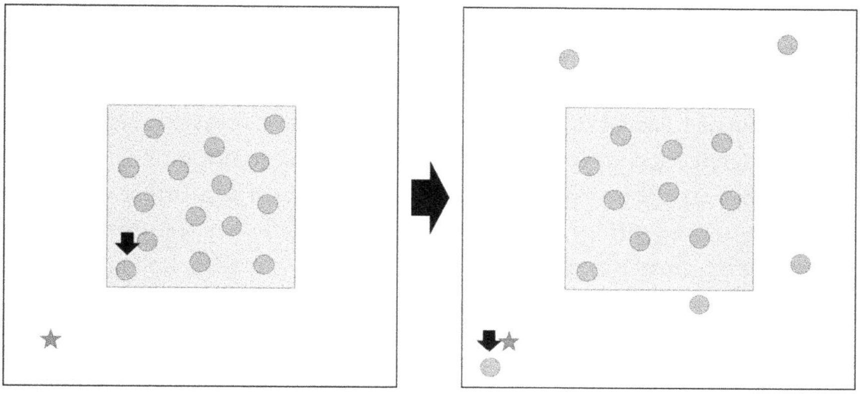

FIGURE 7.7 Effect of Cauchy distribution-based arrangement.

by adjusting the parameter of Cauchy distribution. Unfortunately, there is no concrete guideline that how many particles are placed outside of the search range. We will make several trials to get a better solution. This Cauchy distribution and the rearrangement described later guarantee the discovery of a solution even when no solution exists in the search range.

The allocation based on the Cauchy distribution was realised by generating random numbers based on the Cauchy distribution (hereinafter "Cauchy random numbers"). The Cauchy random number x_c can be generated for a uniform random number x by Eq. (7.6), where μ is the median and γ is the scale. In this program, we first generated the values of the position coordinates by a uniform random number within the set search range and then transformed them into a Cauchy distribution according to this formula to realise placement by a Cauchy random number.

$$x_c = \mu + \gamma \tan\left(\pi\left(x - \frac{1}{2}\right)\right) \tag{7.6}$$

7.5.3 Dynamic Rearrangement of Particle Based on the Average Speed of Particle Group and Optimisation Location

Due to its convergence property, PSO will converge to an inferior solution early if there is no true solution within the range of particle placement, and the search will not proceed. To avoid this, we have already mentioned the need to rearrange the particles. The simplest method is to repeat initialisation and execution until a true solution is found, but this is wasteful.

While repeating this cycle, the search range may be changed. Therefore, our program is designed to perform relocation according to the velocity of the particles existing in the county and the current best position. This allows us to achieve convergence to an inferior solution. This enables automatic detection of convergence to an inferior solution and relocation of particles. Since the problem we are dealing with is a linear system of simultaneous equations, we believe that there is only one solution and that we can get closer to the true solution by focusing our search near the inferior solution, so we have made relocation to be performed around the current best position. The size of the search *range* is set as shown in Eq. (7.7) using the evaluation value *score*. This is based on the idea that since the evaluation value *score* is the sum of the squares of the errors on the left and right sides of the equation, the maximum error per variable that can be

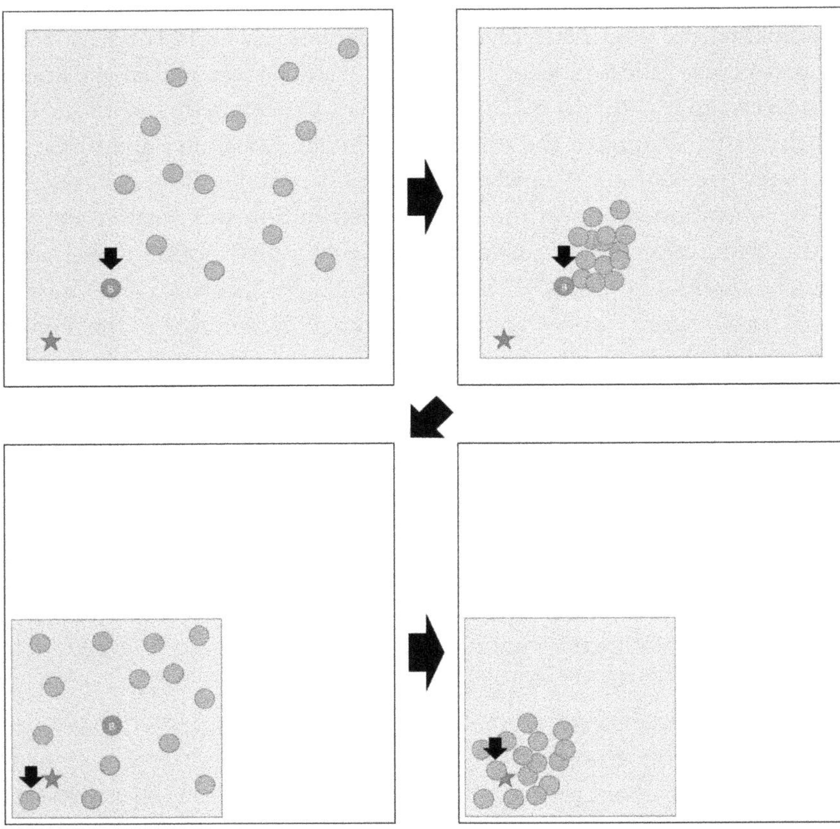

FIGURE 7.8 Rearrangement of particles based on proposed method.

assumed can be obtained by using the number of dimensions n (number of equations = number of variables) as in Eq. (7.7). This allows us to place particles in the range where the true solution is expected to exist, centered on the current best position.

$$range = \frac{1}{n}\sqrt{\frac{score}{n}} \tag{7.7}$$

Figure 7.8 shows an image of the proposed relocation method. At the upper left of the figure, the particles converge to the current best position because there is no true solution in the range of particle placement, but as the convergence progresses, the velocity of particles in the space decreases, and rearrangement occurs. The rearrangement places the particles in the range of the maximum error assumed to be centered on the current best position, so that the arrangement is as shown in the lower right and the true solution can be approached.

7.6 EXPERIMENT AND EVALUATION

7.6.1 Numerical Computation Based on the Proposed Method

In this section, we describe the results of numerical calculations of linear simultaneous equations using the program implemented in this chapter. First, we introduce the numerical computation of binary linear simultaneous equations by PSO; in this chapter, we have the following binary linear simultaneous equations (Eq. (7.8)).

$$\begin{aligned} x + y &= 3 \\ 2x + 5y &= 9 \end{aligned} \tag{7.8}$$

Figure 7.9 plots the movement of particles on a graph. The points on the graph are the particles, and it can be seen that the particles, which are initially distributed evenly over the entire area, gradually converge to the point where the straight lines intersect (the true solution) as the search progresses. The numerical search was completed in 30 steps, with an error of less than 10^{-10}.

7.6.2 Comparison of Ordinary PSO and Proposed Method

In order to demonstrate the effectiveness of the proposed method, which introduces the Cauchy distribution and dynamic rearrangement, a comparison with ordinary PSO is made. In this study, the same 100 linear simultaneous equations were solved using each method with 100 particles. Figure 7.10 shows how the solution improves as the number of iterations

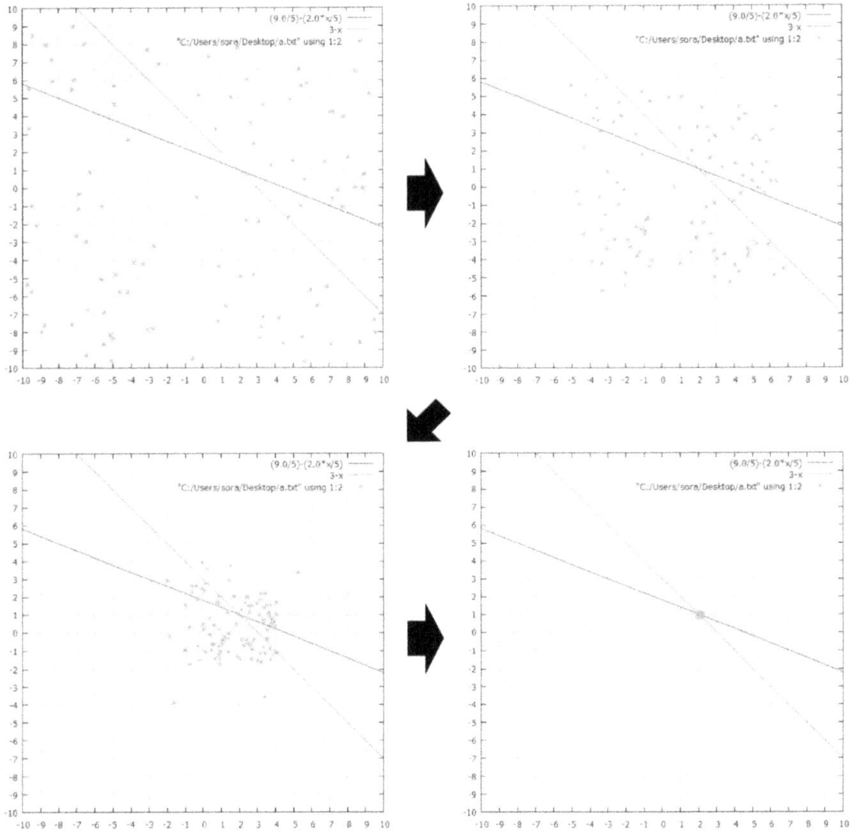

FIGURE 7.9 Numerical computation of Eq. (7.8).

increases. The dashed blue line represents the solution by the normal PSO method, which initially improves at the same rate as that by the proposed method, but after a certain value, the solution hardly improves at all. On the other hand, the proposed method shown as the orange line continuously improves the solution as the search progresses, although the rate of improvement decreases. This indicates that the proposed method is effective to a certain extent.

7.6.3 Assessment in Higher-Order Dimensions

Next, we evaluate the performance in high dimensions. First, Table 7.3 shows the average error evaluation in low dimensions. The table shows the error averages of the true solution after 1000 iterations with 100 particles.

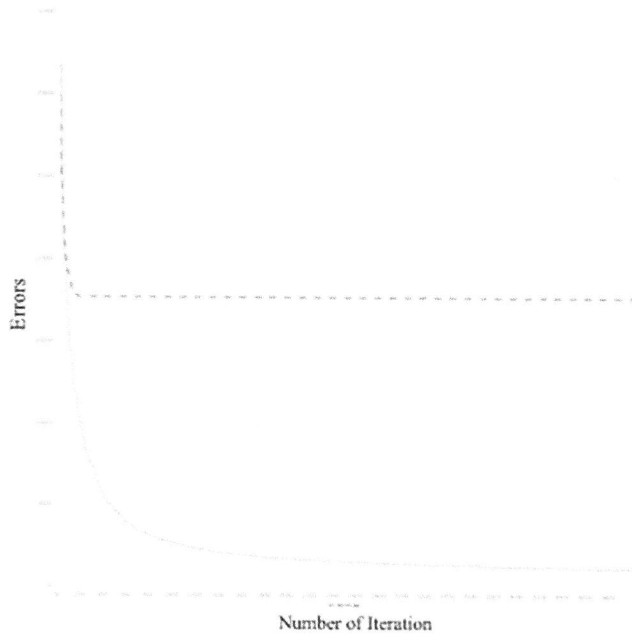

FIGURE 7.10 Comparison of ordinary PSO (dashed blue line) and proposed method (orange line).

TABLE 7.3 Average Error Evaluation in Lower Dimension

Dimension	Average Error
1	0.000
2	0.000
3	0.001
4	0.002
5	0.008

As can be seen, in low dimensions, the solution can be obtained with relatively high accuracy even after 1000 iterations.

Figure 7.11 shows the performance evaluation in higher dimensions. The figure also shows the error averaged over 1000 iterations with the number of particles set to 100. It can be seen that the error increases in proportion to the number of dimensions n. Since the convergence speed of PSO decreases as the search progresses, the number of iterations required to obtain a highly accurate solution is expected to make an even greater difference.

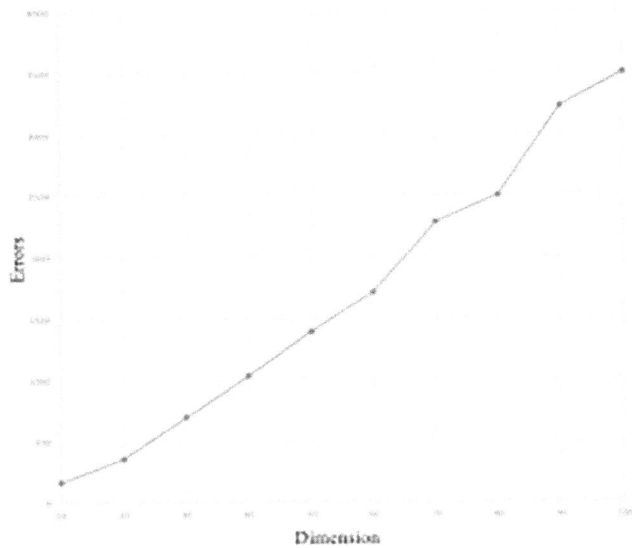

FIGURE 7.11 Efficiency in higher order dimension (1).

In addition, Figure 7.12 shows the number of iterations required to reach the solution for each dimension, assuming that the solution accuracy (allowable error for one variable) is 1.0. The number of particles is 100 as before. The graph shows that the number of iterations increases nearly twice as much for each additional five dimensions, indicating that the overall trend is toward an exponential function.

7.6.4 Relation of the Number of Particles and Iteration Count

We will discuss how the performance of PSO changes as the number of particles increases. Figure 7.13 shows the number of iterations and computation time for each number of particles when solving a 30 original linear system of simultaneous equations with a solution accuracy of 1.0. The solid line in the figure represents the number of iterations, and the dashed line represents the computation time. It can be seen that the number of iterations decreases as the number of particles increases. The sharp decrease in the number of iterations from 100 to 200 particles is thought to be due to the fact that the number of iterations in 30 dimensions is exactly at the boundary of the number of particles for which PSO can demonstrate its full performance. In addition, while the number of iterations is decreasing, the computation time continues to increase after dropping once at the

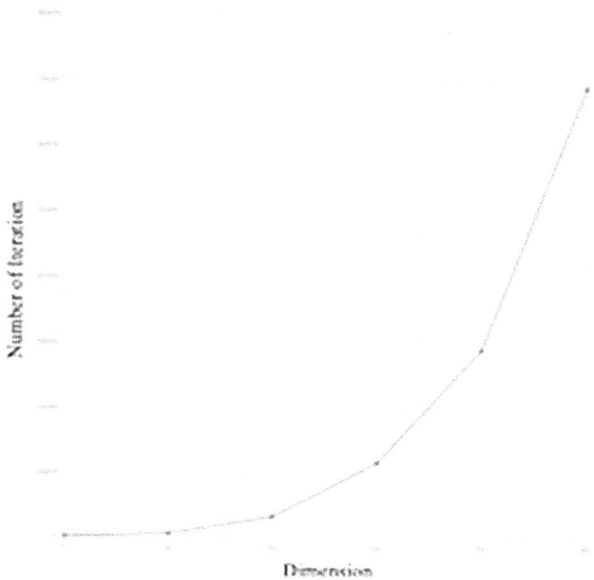

FIGURE 7.12 Efficiency in higher order dimension (2).

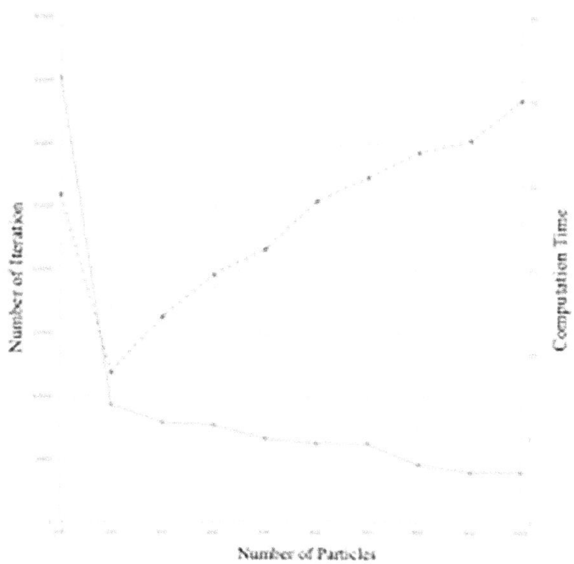

FIGURE 7.13 Relation of the number of particles and computation time for each number of particles.

number of particles of 200. This is because the increase in the computation time per step due to the increase in the number of particles is larger than the decrease in the number of iterations. In other words, it is important to assign the optimal number of particles to each problem in order to run PSO at high speed.

7.7 DISCUSSION AND FUTURE WORK

Numerical computation of linear simultaneous equations by MAS was found to be feasible by using PSO. This was achieved by designing the adaptivity function so that the evaluated value of the particle position is the sum of the squares of the errors between the left-hand side and right-hand side values of the linear simultaneous equations. Since the usual PSO has poor search performance in high-dimensional spaces, it was necessary to change the rearrangement and arrangement of particles to one based on the Cauchy distribution in order to support as high a dimension as possible while guaranteeing the discovery of solutions. However, even with the proposed method, problems of up to about 40 dimensions could be solved in a realistic amount of time. For example, solving a 40-dimension problem required about 10 minutes. This may be related to the fact that a sufficient number of particles has not been obtained. First, we define a sufficient number of particles as the number of particles that can guarantee the discovery of a solution without compromising the original performance of PSO. For example, in the case of a two-dimensional search space, if particles are placed in the four corners of the space as shown in Figure 7.14, one of the four particles will be the particle closest to the solution (if the solution is in the center or on the central axis, the closest particle is not determined, but the search is conducted in the same way), and the other particles will move closer to that particle. Then, in the process of moving, a position closer to the solution is found. By repeating this process, a solution can be found.

Here, the angles of space are represented by the combination of the minimum and maximum values of each axis. In the above two-dimensional case, there are four angles: (*min, min*), (*min, max*), (*max, min*) and (*max, max*). In other words, the number of angles in an n-dimensional space is 2^n, and we can consider that 2^n particles are necessary in the case of ordinary PSO to ensure the discovery of a solution. Therefore, the sufficient number of particles in n-dimensional space is considered to be more than 2^n. It is practically impossible to obtain such a large number of resources with current computers, and this is thought to be the reason for the poor performance in higher dimensions.

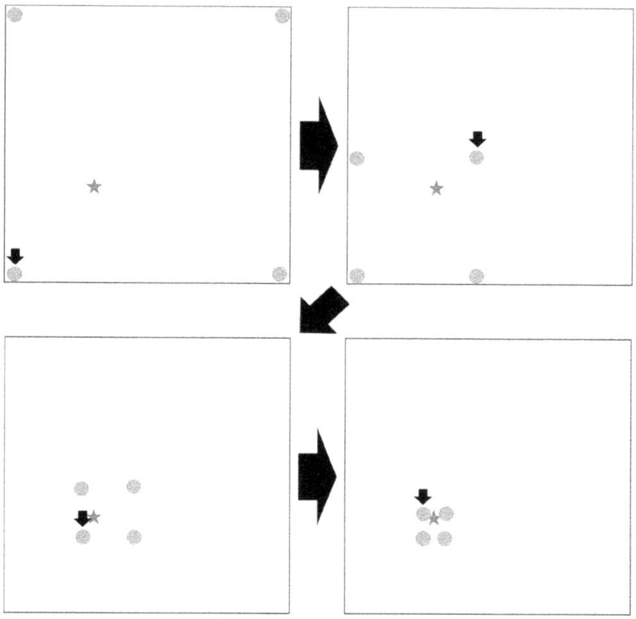

FIGURE 7.14 Searching from 4 corners of a 2-dimensional space.

As mentioned earlier, the performance degradation in higher dimensional space is a major issue. The program developed in this study can solve linear systems in a reasonable time only up to about 40 dimensions, and beyond that, it takes an enormous amount of time. However, the linear simultaneous equations actually handled in the engineering field have thousands or tens of thousands of dimensions, and our program is still insufficient to handle actual problems. This is a problem that PSO itself has, regardless of linear simultaneous equations, and it is difficult to improve it. One way to improve this problem is, for example, to solve 100 linear simultaneous equations by dividing them into 100 problems of 10 elements each, or to focus on some property of linear simultaneous equations and solve them with PSO that provides knowledge. In the case of numerical computation of linear simultaneous equations by MAS, it may be possible to use coefficients as agents, and to perform operations while exchanging information with each other so that the agents of diagonal components are set to 1 and the agents of other components to 0. In addition, since MAS has good compatibility with parallel processing, we believe that the speed can be improved by introducing parallel processing.

7.8 CONCLUSION

Numerical computation of the simultaneous equations by MAS was found to be feasible by using PSO. This can be achieved by designing the adaptivity function so that the evaluated value of the particle position is the sum of the squares of the errors in the left-hand and right-hand sides of the simultaneous equations. However, in ordinary PSO, once convergence to an inferior solution is achieved, the search cannot proceed any further, and the discovery of the true solution cannot be guaranteed. In order to avoid convergence to an inferior solution and guarantee the discovery of a true solution, it was necessary to rearrange the particles. Since PSO does not perform as well as it should in high dimensions, and search efficiency decreases, we improved the particle placement and rearrangement methods. As a result, we were able to solve high-dimensional problems that cannot be solved by PSO. For the placement of particles, we used the Cauchy distribution, which has been proposed in previous research, to achieve both global search and intensive search, and to guarantee the discovery of a solution. In addition, by controlling the rearrangement according to the speed of the particles and the current best position, we were able to speed up the convergence speed while guaranteeing the discovery of the true solution.

However, even with the proposed method, the problem can be solved in a realistic time up to about 40 dimensions, and it is difficult to solve linear simultaneous equations with thousands or tens of thousands of elements, such as those used in actual engineering fields. As mentioned earlier, the number of particles required for the PSO to achieve its full performance is 2^n or more, and the higher the dimension, the more the current computer cannot handle the problem at all. In order to deal with higher dimensional problems, we can consider methods such as preprocessing the given problem to make it lower dimensional or dividing the problem into smaller pieces and solving them separately, but this is a subject for future research.

7.9 ACKNOWLEDGEMENT

The authors would like to express their sincere thanks to some undergraduate students of our research group who contributed the development of programs to execute numerical experiment.

REFERENCES

Ajith Abraham, A., Grosan, C., & Ramos, V. (2006). *Swarm intelligence and data mining* (Vol. 34, Studies in computational intelligence). Springer.

Czarnowski, I., Jędrzejowicz, P., & Kacprzyk, J. (Eds.). (2013). *Agent-based optimization* (Studies in computational intelligence). Springer.

Feller, W. (1971). *An introduction to probability theory and its applications.* John Wiley & Sons.

Hassanien, A. B., & Emary, E. (2015). *Swarm intelligence: Principles, advances, and applications.* CRC Press.

Hoffmeister, F., & Bäck, T. (1991). Genetic algorithms and evolution strategies: Similarities and differences. In H. P. Schwefel & R. Männer (Eds.), *Parallel problem solving from nature* (Vol. 496). Lecture Notes in Computer Science. Springer.

Kaji, T. (2014). *Scatter search Cauchy particle swarm optimization for multimodal functions.* www.jstage.jst.go.jp/article/jima/64/4/64_510/_pdf

Parsopoulos, K. E., & Vrahatis, M. N. (2004). On the computation of all global minimizers through particle swarm optimization. *IEEE Transactions on Evolutionary Computation*, *8*(3), 211–224. https://doi.org/10.1109/TEVC.2004.826076

Russel, S., & Norvig, P. (2003). *Artificial intelligence: A modern approach.* Prentice Hall.

Shimizu, Y., & Haga, H. (2018). Solving partial differential equations using multi-agent modelling. *AIP conference proceedings MathzTech2018*, 060018-1-060018-11.

Simon Fraser University. (2013). *Virtual library of simulation experiments: Test functions and datasets.* www.sfu.ca/~ssurjano/index.html

Wilensky, U., & Rand, W. (2015). *An introduction to agent-based modeling: Modeling natural, social, and engineered complex systems with NetLogo.* MIT Press.

CHAPTER 8

Tracking Abrupt Changes in Physical and Engineering Systems Using Recursive System Identification

Noor Atinah Ahmad and Wei Shan Ng

8.1 INTRODUCTION

The identification of nonlinear dynamics is the process of determining and characterising the underlying nonlinear relationships and behaviour of a dynamical system from observed data. This task is crucial across various fields, including physics, engineering, biology and economics, where real-world systems often exhibit nonlinear behaviour that cannot be adequately captured by simple linear models. The process typically starts with the collection of time-series data that represents the behaviour of the system of interest over time. This data could be obtained through experiments, observations or measurements. Raw data often contains noise and artifacts. Data preprocessing techniques, such as filtering and detrending, are applied to clean the data and make it suitable for analysis. The next step is to select an appropriate mathematical model that captures the nonlinear dynamics of

DOI: 10.1201/9781032678573-8

the system. This may involve considering various candidate models, including polynomial models, neural networks or more specialised models tailored to the specific system under investigation. Parameters of the chosen model are then estimated from the observed data. This can be done using optimisation techniques like least squares or maximum likelihood estimation. The goal is to find the parameter values that best fit the observed data.

Recursive system identification is designed to work with streaming or continuously updated data, making it suitable for real-time analysis and monitoring of evolving systems. It can adapt to incoming data without requiring retraining on the entire dataset. This is achieved using a recursive approach, where new data points are used to update the system's model incrementally. This process can be computationally efficient and avoids the need to recompute the model parameters from scratch. Recursive system identification is useful in applications where real-time or near-real-time insights into the system's dynamics are crucial. It can adapt to changing system dynamics over time. As new data reveals modifications in the system's behaviour, recursive system identification can update its model to account for these changes.

8.2 BASIC CONCEPTS

To understand the mathematical and algorithmic development of recursive system identification, we begin by outlining the basic concepts of a dynamical system and the identification of system parameters from data.

8.2.1 Dynamical System

A continuous dynamical system can be described mathematically in the form

$$\frac{dX}{dt} = F\left(X\left(t\right), \alpha\right), \tag{8.1}$$

where $X\left(t\right) \in R^n$ is the state of the dynamical system at time t, the function F represents the mathematical expression for the dynamics and $\alpha \in R^p$ is a vector of p parameters of the system.

8.2.1.1 Example 8.1: The Lorenz System

The Lorenz system is a well-studied, and highly simplified, conceptual model for atmospheric circulation. The state vector of the system $X\left(t\right) = \left(x\left(t\right), y\left(t\right), z\left(t\right)\right)^T$ consists of three quantities with respect to time: x is proportional to the rate of convection, y to the horizontal temperature variation and z to the vertical temperature variation. The dynamical system model is a system of three first order differential equations given by

$$\frac{dx}{dt} = \sigma(y - x), \tag{8.2a}$$

$$\frac{dy}{dt} = x(\rho - z) - y, \tag{8.2b}$$

$$\frac{dz}{dt} = xy - \beta z, \tag{8.2c}$$

where the parameter σ corresponds to the Prandtl number, ρ represents the heating of the atmosphere corresponding to the Rayleigh number and β refers to the aspect ratio. If we set $\sigma = 10$ and $\beta = 8/3$, distinct dynamics can be observed for different values of ρ. Figure 8.1 exhibits three distinct dynamics that arise from three different values: $\rho = 28, 100$, and 0.8. For $\rho = 28$, the system produces a chaotic solution; an example of the time series for y (see Figure 8.1a) shows the unpredictable nature of the solution over time. The associated phase portrait (see Figure 8.1b) reveals the infamous Lorenz attractor; the "butterfly effect" that represents the phase space in which different initial chaotic conditions generate distinct trajectories that never repeat. When $\rho = 100$, the system exhibits a periodic solution (see Figure 8.1c) with the associated phase portrait given in Figure 8.1d portraying a solution trajectory going through a double loop. For $\rho = 0.8$, an asymptotically stable solution is observed (see Figure 8.1e) with the associated phase portrait provided in Figure 8.1f.

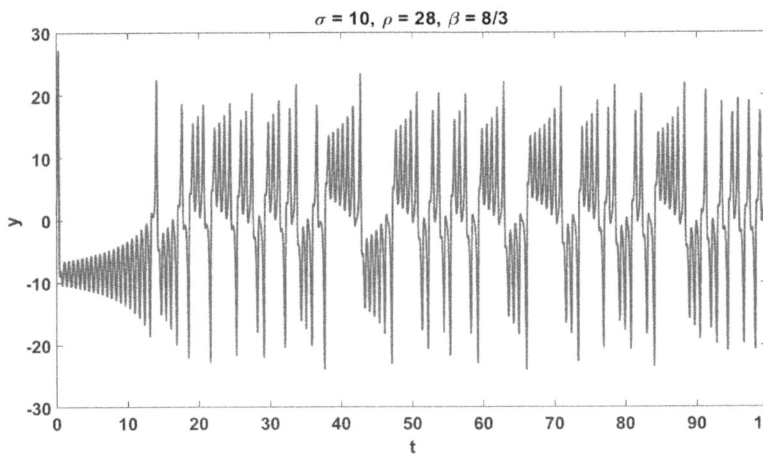

(a)

FIGURE 8.1 Dynamics of the Lorenz system for different value of the parameter ρ_2 while keeping $\sigma_1 = 10$, $\beta_3 = 8/3$. (a), (c) and (e) Time series of state variable y for $\rho_2 = 28,100,0.8$, respectively. (b), (d) and (f) Phase portraits of the system for $\rho_2 = 28,100,0.8$, respectively. (Continued)

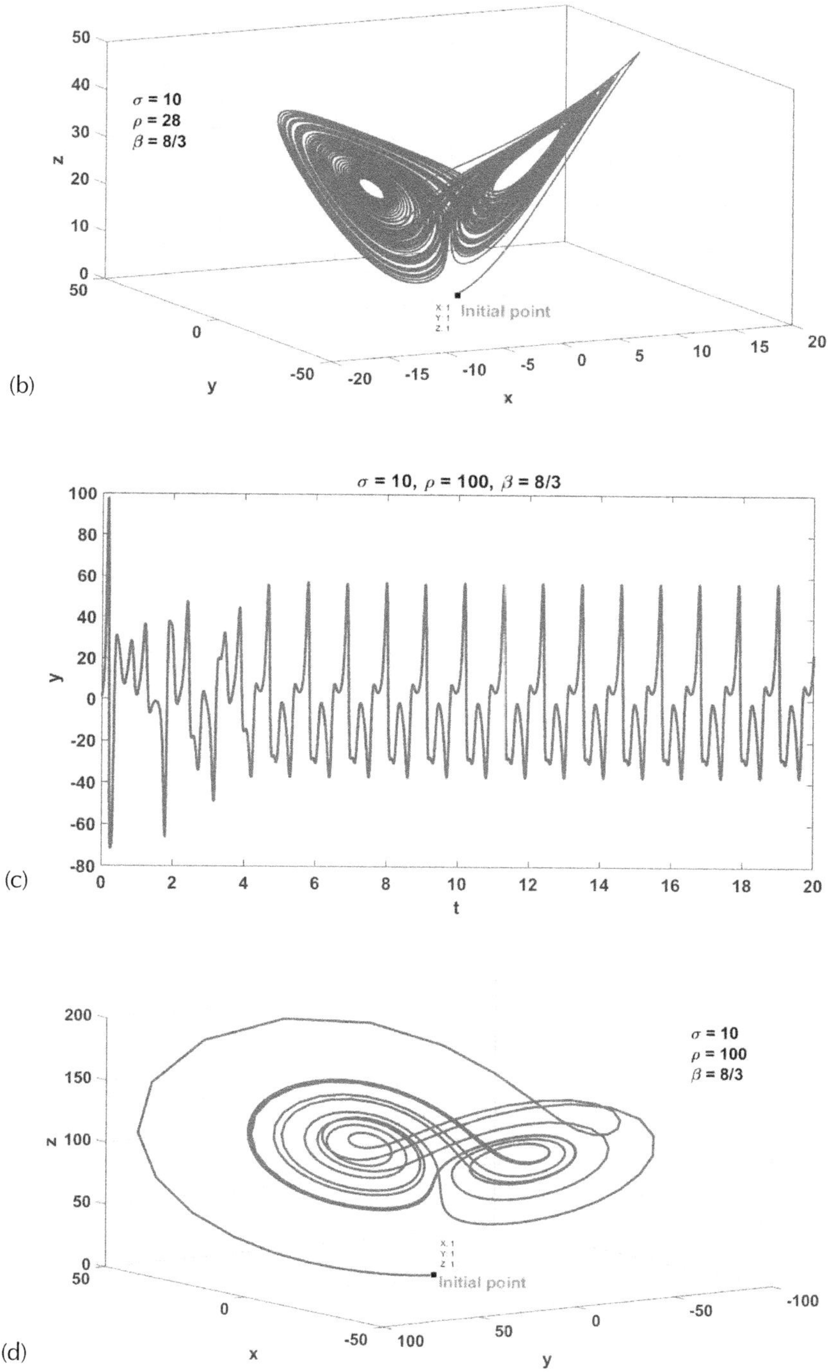

FIGURE 8.1 (Continued)

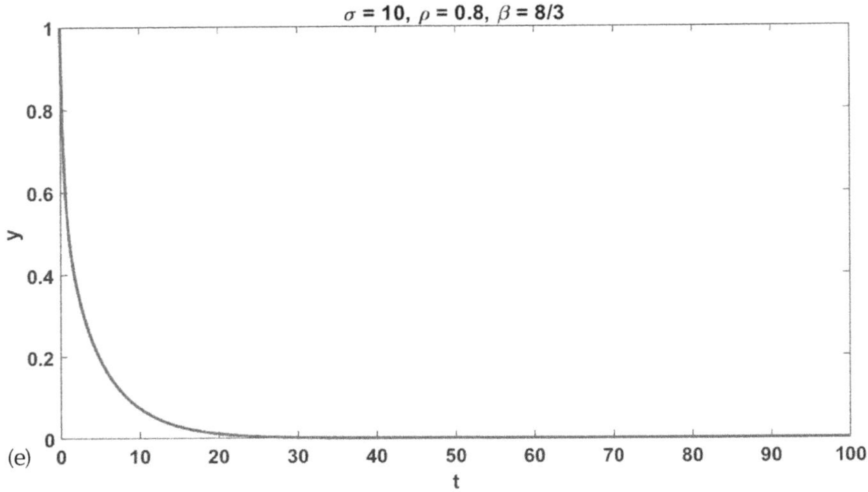

(e)

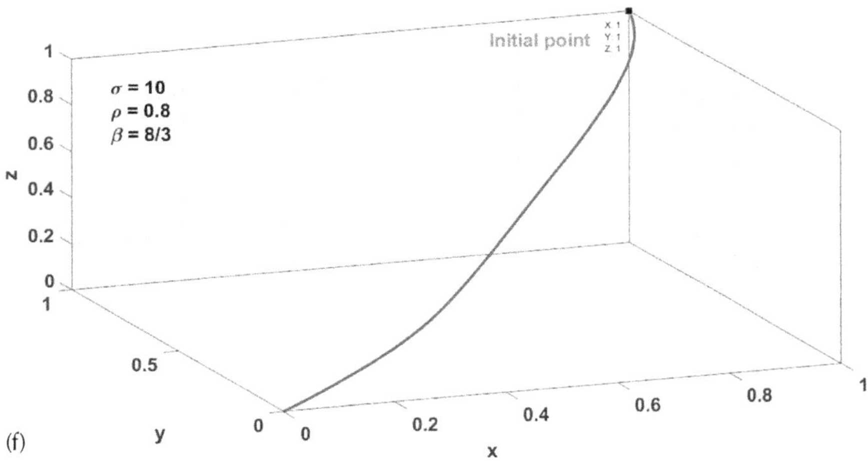

(f)

FIGURE 8.1 (Continued)

8.2.1.2 Example 8.2: Van der Pol Oscillator

The Van der Pol equation was first introduced by the Dutch electrical engineer Balthazar Van der Pol in 1927 to describe oscillations in a vacuum tube electrical circuit. Given $x(t)$ as the dependent variable representing the displacement or voltage, t is time, the unforced Van der Pol equation is given by

$$\frac{d^2x}{dt^2}+\mu\left(1-x^2\right)\frac{dx}{dt}+x=0, \tag{8.3}$$

where the positive number μ parameterises the nonlinear damping term. By letting $y = \frac{dx}{dt}$, the above equation can be written in the form of a system of first-order differential equations given by

$$\frac{dx}{dt} = y, \tag{8.4a}$$

$$\frac{dy}{dt} = -\mu\left(1 - x^2\right)y - x. \tag{8.4b}$$

The dynamics of the Van der Pol system are characterised by the relaxation oscillations, or self-sustained oscillations that are viewable in the phase space as limit cycles. Figure 8.2a–f displays the evolution of the time series and limit cycles as the value of μ changes from 0.1 to 2.0, and finally to 4.0.

8.2.2 Learning Dynamical System from Data

Traditional methods for identifying nonlinear dynamics often require explicit knowledge of the system's governing equations. However, in many real-world scenarios, obtaining accurate and complete models of complex nonlinear systems is challenging. Assuming a dynamical system model

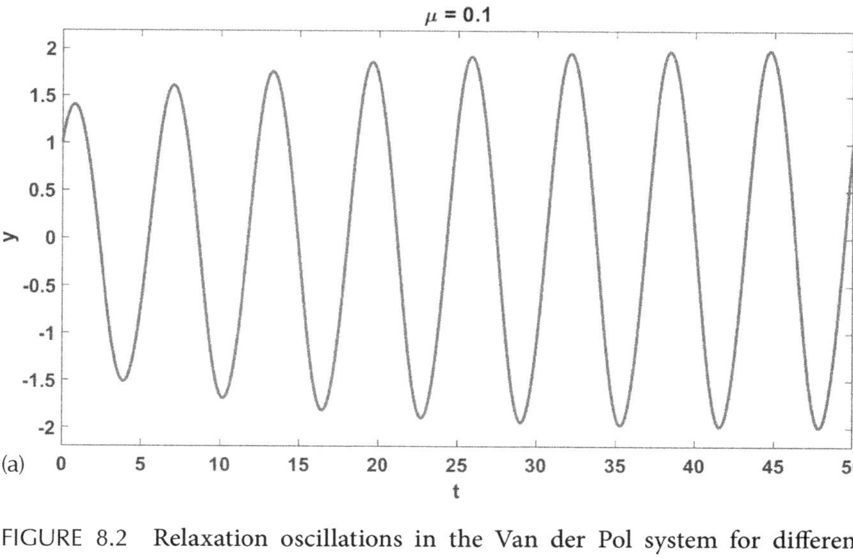

(a)

FIGURE 8.2 Relaxation oscillations in the Van der Pol system for different value of the parameter μ. (a), (c) and (e) Time series of state variable y for $\mu = 0.1, 2.0, 4.0$, respectively. (b), (d) and (f) Phase portraits of the system for $\mu = 0.1, 2.0, 4.0$, respectively. (Continued)

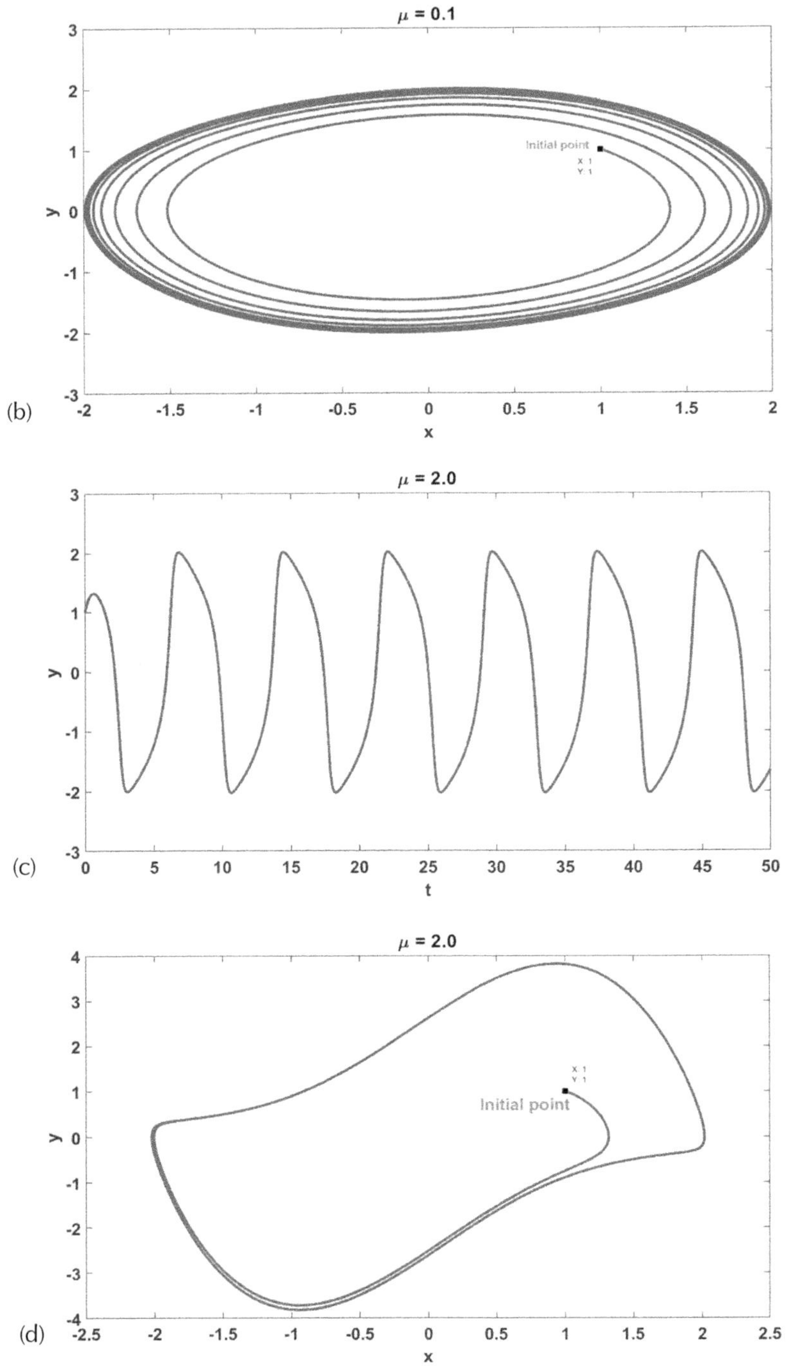

FIGURE 8.2 (Continued)

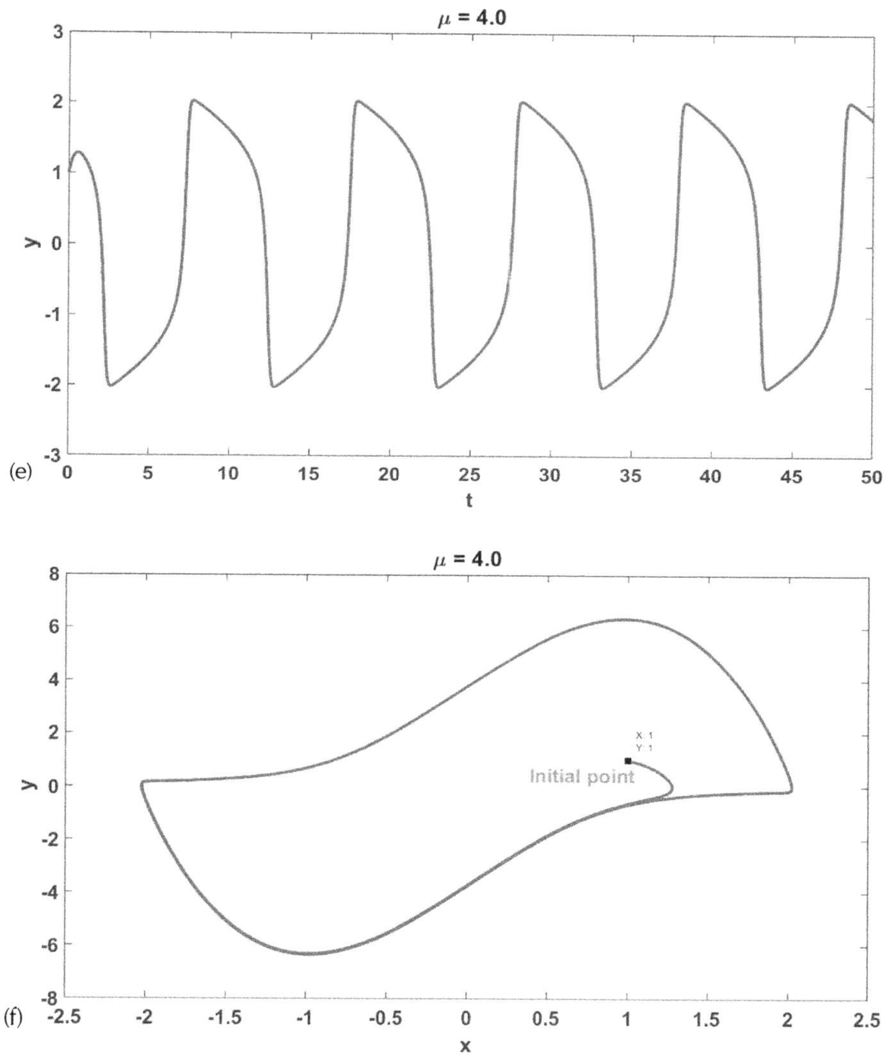

FIGURE 8.2 (Continued)

in Eq. (8.1), if data for the state-variables $X(t)$ can be measured, the unknown dynamics described by the vector field F can be approximated using regression.

Let $F = [f_1, f_2, \ldots, f_n]^T$ where $f_1, f_2, \ldots, f_n$ belong to some function space V whose basis functions are $\phi_1, \phi_2, \ldots, \phi_k$, and k need not be finite. Then for each $i = 1, 2, \ldots, n$, f_i assumes a linear expansion of the form

$$F(X(t),\alpha) = \begin{bmatrix} f_1(X(t),\alpha) \\ f_2(X(t),\alpha) \\ \vdots \\ f_n(X(t),\alpha) \end{bmatrix} = \begin{bmatrix} \alpha_{11} & \alpha_{12} & \cdots & \alpha_{1k} \\ \alpha_{21} & \alpha_{22} & \cdots & \alpha_{2k} \\ \vdots & \vdots & \ddots & \vdots \\ \alpha_{n1} & \alpha_{n2} & \cdots & \alpha_{nk} \end{bmatrix} \begin{bmatrix} \phi_1(X(t)) \\ \phi_2(X(t)) \\ \vdots \\ \phi_k(X(t)) \end{bmatrix} = \alpha\varphi(t)$$

$$\text{with } \alpha = \begin{bmatrix} \alpha_{11} & \alpha_{12} & \cdots & \alpha_{1k} \\ \alpha_{21} & \alpha_{22} & \cdots & \alpha_{2k} \\ \vdots & \vdots & \ddots & \vdots \\ \alpha_{n1} & \alpha_{n2} & \cdots & \alpha_{nk} \end{bmatrix}, \quad \varphi(t) = \begin{bmatrix} \phi_1(X(t)) \\ \phi_2(X(t)) \\ \vdots \\ \phi_k(X(t)) \end{bmatrix}.$$

FIGURE 8.3 The vector form for F written in matrix-vector product.

$$f_i\left(X(t),\alpha\right) = \alpha_{i1}\phi_1\left(X(t)\right) + \alpha_{i2}\phi_2\left(X(t)\right) + \cdots + \alpha_{ik}\phi_k\left(X(t)\right), \quad (8.5)$$

where $\alpha_{i1}, \alpha_{i2}, \ldots, \alpha_{ik}$ are parameters to be determined. The form given in Eq. (8.5) allows us to write F as a matrix-vector product with details given in Figure 8.3.

The matrix-vector product form of the dynamics vector field F provides a linear model for F in the form $\alpha\varphi(t)$ in the space V. A mathematically tractable way to determine α from data is to use regression. Given some observed measurements of the time derivative, $\dot{X}(t) = \frac{dX(t)}{dt}$, the best-fit model can be obtained by finding the parameter matrix α that minimises the distance between the observed function $\dot{X}(t)$ and its model $\alpha\varphi(t)$. Mathematically, this is achieved by solving the following regression problem

$$\alpha_{opt} = \arg\min \| \dot{X}(t) - \alpha\varphi(t) \|, \quad (8.6)$$

where α_{opt} is the optimum parameter matrix.

To determine α_{opt} from data, we either measure the derivative $\frac{dX(t)}{dt}$ or approximate it numerically from $X(t)$. The data are sampled at several times $t_1, t_2, \ldots, t_m$ and arranged into a data matrix Y_m. Using sampled data of the state $X(t)$, basis functions $\phi_1, \phi_2, \ldots, \phi_k$ (assumed known) are evaluated and arranged in the respective columns of the basis matrix Φ_m. The detailed construction of matrices Y_m and Φ_m is shown in Figure 8.4. With these constructions, the regression problem in Eq. (8.6) can be recast as a minimisation problem of the form

$$Y_m = \begin{bmatrix} \dot{X}^T(t_1) \\ \dot{X}^T(t_2) \\ \vdots \\ \dot{X}^T(t_m) \end{bmatrix} = \begin{bmatrix} \dot{x}_1(t_1) & \dot{x}_2(t_1) & \cdots & \dot{x}_n(t_1) \\ \dot{x}_1(t_2) & \dot{x}_2(t_2) & \cdots & \dot{x}_n(t_2) \\ \vdots & \vdots & \ddots & \vdots \\ \dot{x}_1(t_m) & \dot{x}_2(t_m) & \cdots & \dot{x}_n(t_m) \end{bmatrix}$$

$$\Phi_m = \begin{bmatrix} \varphi^T(t_1) \\ \varphi^T(t_2) \\ \vdots \\ \varphi^T(t_m) \end{bmatrix} = \begin{bmatrix} \phi_1(\mathbf{x}(t_1)) & \phi_2(\mathbf{x}(t_1)) & \cdots & \phi_k(\mathbf{x}(t_1)) \\ \phi_1(\mathbf{x}(t_2)) & \phi_2(\mathbf{x}(t_2)) & \cdots & \phi_k(\mathbf{x}(t_2)) \\ \vdots & \vdots & \ddots & \vdots \\ \phi_1(\mathbf{x}(t_m)) & \phi_2(\mathbf{x}(t_m)) & \cdots & \phi_k(\mathbf{x}(t_m)) \end{bmatrix}$$

FIGURE 8.4 The layout of matrices Y_m and Φ_m.

$$\alpha_{opt} = \arg\min \| Y_m^T - \alpha \Phi_m^T \|_F^2, \tag{8.7}$$

where $\|\cdot\|_F$ refers to the Frobenius norm. It follows that the minimum-norm solution of Eq. (8.7) is given by

$$\alpha_{opt} = Y_m^T \Phi_m \left(\Phi_m^T \Phi_m \right)^{-1}. \tag{8.8}$$

8.2.2.1 Example 8.3

Consider a dictionary of basis functions for the second-degree polynomial:

$$\phi_0 = 1,\ \phi_1 = x_1, \phi_2 = x_2, \phi_3 = x_3, \phi_4 = (x_1)^2, \phi_5 = x_1 x_2, \phi_6 = x_1 x_3,$$
$$\phi_7 = (x_2)^2, \phi_8 = x_2 x_3, \phi_9 = (x_3)^2.$$

Based on this dictionary, the Lorenz system from Example 8.1 can be written as follows:

$$\frac{dx_1}{dt} = \alpha_{11}\phi_1 + \alpha_{12}\phi_2, \tag{8.9a}$$

$$\frac{dx_2}{dt} = \alpha_{21}\phi_1 + \alpha_{22}\phi_2 + \alpha_{26}\phi_6, \tag{8.9b}$$

$$\frac{dx_3}{dt} = \alpha_{33}\phi_3 + \alpha_{35}\phi_5. \tag{8.9c}$$

Here, we identify the state variables as $x_1 = x, x_2 = y, x_3 = z$, and the parameters as $\alpha_{11} = -\sigma, \alpha_{12} = \sigma, \alpha_{21} = \rho, \alpha_{22} = -1, \alpha_{26} = -1$, $\alpha_{33} = -\beta$ and $\alpha_{35} = 1$.

8.2.2.2 Example 8.4

Consider a dictionary of basis functions for the third-degree polynomial:

$$\phi_0 = 1, \phi_1 = x_1, \phi_2 = x_2, \phi_3 = x_1^2, \phi_4 = x_1 x_2, \phi_5 = x_2^2, \phi_6 = x_1^3,$$
$$\phi_7 = x_1^2 x_2, \phi_8 = x_2^3, \phi_9 = x_1 x_2^2.$$

Based on this dictionary, the Van der Pol system from Example 8.2 can be written as follows:

$$\frac{dx_1}{dt} = \alpha_{12}\phi_2, \tag{8.10a}$$

$$\frac{dx_2}{dt} = \alpha_{21}\phi_1 + \alpha_{22}\phi_2 + \alpha_{27}\phi_7. \tag{8.10b}$$

Here, we identify the state variables as $x_1 = x, x_2 = y = \frac{dx}{dt}$, and the parameters as $\alpha_{12} = 1, \alpha_{21} = -1, \alpha_{22} = -\mu$ and $\alpha_{27} = \mu$.

8.3 RECURSIVE SYSTEM IDENTIFICATION

In the real-time change detection of physical and engineering systems, data is collected continuously to monitor the system, and to detect and respond to changes as they occur. The ability for the system to react immediately to changes in data or events is crucial for system recovery. From the perspective of dynamical system models, changes in data or events signify changes in the system dynamics, and for the effective control of the system, it is important to implement a mechanism that can rapidly characterise and compensate for the new dynamics. This calls for a methodology that detects changes in system's parameters using a limited amount of data and responds in as little time as possible.

The mathematical formulation provided in section 8.2.2 requires the computation of the optimum model parameters given by Eq. (8.8) that relies on the knowledge of the data matrix Y_m and basis matrix Φ_m. However, in real-time processing, these matrices are incomplete and are dynamically updated with every incoming data. In this section we describe the recursive updating of Y_m and Φ_m, and how the process leads to the recursive updating of the optimum system parameter matrix, α_{opt}. A change detection mechanism is then incorporated into the recursive updating process to detect changes and promptly initiate the recovery of system's parameters.

8.3.1 Recursive Updating of the Optimum Parameter Matrix

Suppose data for Y_m and Φ_m are available up to time t_m. With the arrival of data $\left\{X\left(t_{m+1}\right), \dot{X}\left(t_{m+1}\right)\right\}$ at the next time sequence t_{m+1}, Φ_m and Y_m are updated to give Φ_{m+1} and Y_{m+1} as shown in Figure 8.5.

According to Eq. (8.8), the optimal parameter matrix at time t_{m+1} is given by

$$\alpha_{opt}^{(m+1)} = Y_{m+1}^T \Phi_{m+1} \left(\Phi_{m+1}^T \Phi_{m+1}\right)^{-1}. \tag{8.11}$$

Now, let $Q_m = Y_m^T \Phi_m$ and $P_m = \Phi_m^T \Phi_m$. Then

$$Q_{m+1} = Y_{m+1}^T \Phi_{m+1} = Q_m + y_{m+1} \varphi_{m+1}^T, \tag{8.12}$$

$$P_{m+1}^{-1} = \left(\Phi_{m+1}^T \Phi_{m+1}\right)^{-1} = \left(P_m + \varphi_{m+1} \varphi_{m+1}^T\right)^{-1}, \tag{8.13}$$

where

$$\varphi_{m+1} = \left[\phi_1\left(X\left(t_{m+1}\right)\right), \phi_2\left(X\left(t_{m+1}\right)\right), \ldots, \phi_k\left(X\left(t_{m+1}\right)\right)\right]^T, \tag{8.14}$$

and

$$y_{m+1} = \left[\dot{x}_1\left(t_{m+1}\right), \dot{x}_2\left(t_{m+1}\right), \ldots, \dot{x}_1\left(t_{m+1}\right)\right]^T. \tag{8.15}$$

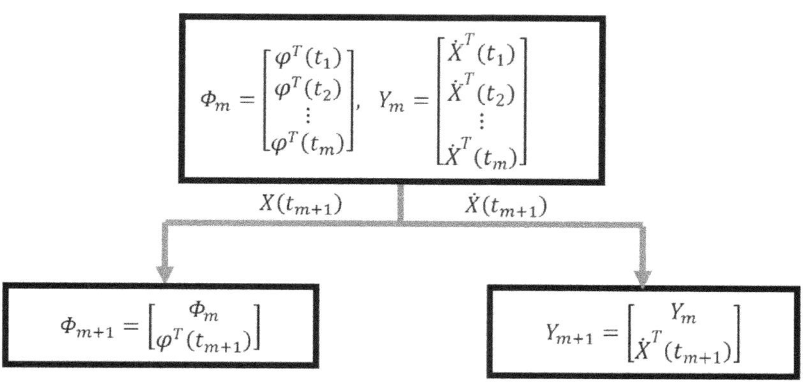

FIGURE 8.5 The update of data matrices from Φ_m and Y_m to Φ_{m+1} and Y_{m+1}, respectively.

Eqs. (8.12) and (8.13) allow for an update of Eq. (8.11) as follows:

$$\alpha_{opt}^{(m+1)} = \left(Q_m + y_{m+1}\varphi_{m+1}^T\right)\left(P_m + \varphi_{m+1}\varphi_{m+1}^T\right)^{-1}. \tag{8.16}$$

The right-hand side expression in Eq. (8.16) can be computed efficiently by making use of the Sherman-Morrison formula introduced in Sherman and Morrison (1950) and used in a similar context by Zhang et al. (2019). Given an invertible matrix $W + uv^T$, according to the Sherman-Morrison formula, the inverse can be updated recursively using the formula

$$\left(W + uv^T\right)^{-1} = W^{-1} - \frac{W^{-1}uv^TW^{-1}}{1 + v^TW^{-1}u}. \tag{8.17}$$

By expanding $\left(P_m + \varphi_{m+1}\varphi_{m+1}^T\right)^{-1}$ using Eq. (8.17), we obtain

$$\begin{aligned}\left(P_m + \varphi_{m+1}\varphi_{m+1}^T\right)^{-1} &= P_m^{-1} - \frac{P_m^{-1}\varphi_{m+1}\varphi_{m+1}^T P_m^{-1}}{1 + \varphi_{m+1}^T P_m^{-1}\varphi_{m+1}} \\ &= P_m^{-1} - \mu_{m+1}P_m^{-1}\varphi_{m+1}\varphi_{m+1}^T P_m^{-1},\end{aligned} \tag{8.18}$$

where $\mu_{m+1} = \left(1 + \varphi_{m+1}^T P_m^{-1}\varphi_{m+1}\right)^{-1}$. After inserting Eq. (8.18) into Eq. (8.16), we get

$$\begin{aligned}\alpha_{opt}^{(m+1)} &= \alpha_{opt}^{(m)} - \mu_{m+1}Q_m P_m^{-1}\varphi_{m+1}\varphi_{m+1}^T P_m^{-1} + y_{m+1}\varphi_{m+1}^T P_m^{-1} \\ &\quad - \mu_{m+1}y_{m+1}\varphi_{m+1}^T P_m^{-1}\varphi_{m+1}\varphi_{m+1}^T P_m^{-1}.\end{aligned} \tag{8.19}$$

The last two terms in Eq. (8.19) can be simplified to

$$\begin{aligned}&y_{m+1}\varphi_{m+1}^T P_m^{-1} - \mu_{m+1}y_{m+1}\varphi_{m+1}^T P_m^{-1}\varphi_{m+1}\varphi_{m+1}^T P_m^{-1} \\ &= \mu_{m+1}y_{m+1}\left(\mu_{m+1}^{-1} - \varphi_{m+1}^T P_m^{-1}\varphi_{m+1}\right)\varphi_{m+1}^T P_m^{-1} \\ &= \mu_{m+1}y_{m+1}\varphi_{m+1}^T P_m^{-1}.\end{aligned} \tag{8.20}$$

Let $\omega_m = \varphi_{m+1}^T P_m^{-1}$. Then,

$$\begin{aligned}\alpha_{opt}^{(m+1)} &= \alpha_{opt}^{(m)} - \mu_{m+1}\alpha_{opt}^{(m)}\varphi_{m+1}\omega_m + \mu_{m+1}y_{m+1}\omega_m \\ &= \alpha_{opt}^{(m)} + \mu_{m+1}\left(y_{m+1} - \alpha_{opt}^{(m)}\varphi_{m+1}\right)\omega_m \\ &= \alpha_{opt}^{(m)} + \mu_{m+1}\varepsilon_{m+1}\omega_m,\end{aligned} \tag{8.21}$$

TABLE 8.1 Basic Algorithm for Recursive Updating of Optimum Parameter Matrix

Given $P_m^{-1}, \alpha_{opt}^{(m)}$ computed at time t_m,
and with current updates y_{m+1} and φ_{m+1} at time t_{m+1}, compute:
$\omega_m = \varphi_{m+1}^T P_m^{-1}$
$\mu_{m+1} = \left(1 + \omega_m \varphi_{m+1}\right)^{-1}$
$\varepsilon_{m+1} = y_{m+1} - \alpha_{opt}^{(m)} \varphi_{m+1}$
$\alpha_{opt}^{(m+1)} = \alpha_{opt}^{(m)} + \mu_{m+1} \varepsilon_{m+1} \omega_m$

where $\varepsilon_{m+1} = y_{m+1} - \alpha_{opt}^{(m)} \varphi_{m+1}$ may be interpreted as the *a priori* instantaneous error vector in the approximation of the dynamics and ω_m is synonymous to the Kalman gain vector. Eq. (8.21) provides a recursive update of the optimum parameter matrix α_{opt} given the updates y_{m+1} and φ_{m+1} at time t_{m+1}. The basic algorithm for the update is provided in Table 8.1.

The computation in Table 8.1 requires a total of $k^2 + (2n+1)k$ multiplications, whereas to compute $\alpha_{opt}^{(m+1)}$ using Eq. (8.11) will require a total of $k^3 + (m+n+1)k^2 + (m+1)nk$ multiplications. So, for a dictionary of basis function of size k, the complexity of the recursive update is only $O(k^2)$ compared to $O(k^3)$ for the nonrecursive counterpart. The nonrecursive update also suffers from instability with increasing k due to the ill-posedness of the regression problem in Eq. (8.7) when k>>n. Whereas for the recursive update, to mitigate the ill-posedness of the problem, we set $P_0^{-1} = \hat{\sigma} I_k$, where $\hat{\sigma}$ is an estimate of the variance of noise in the data.

8.3.2 Identifying Abrupt Changes: Change Point Detection Using *k*-Nearest Neighbour (*k*-NN) Classifier

When there is a change in the system's parameters, there will be an abrupt change in the orbit in the phase-space, leading to a trajectory of data points that moves away from the initial trajectory. Consider a set of K consecutive points $\left\{X\left(t_{i-+1}\right), X\left(t_{i-K+2}\right), \ldots, X\left(t_{i-1}\right), X\left(t_i\right)\right\}$ on an initial trajectory T_1. If a change in the system's parameter occurs at time t_{i+1}, a new trajectory T_2 will emerge giving rise to a new set of data points $\left\{X\left(t_{i+1}\right), X\left(t_{i+2}\right), \ldots\right\}$ (see illustrations in Figure 8.6). The set of points on T_1 and T_2 can be considered coming from two different classes.

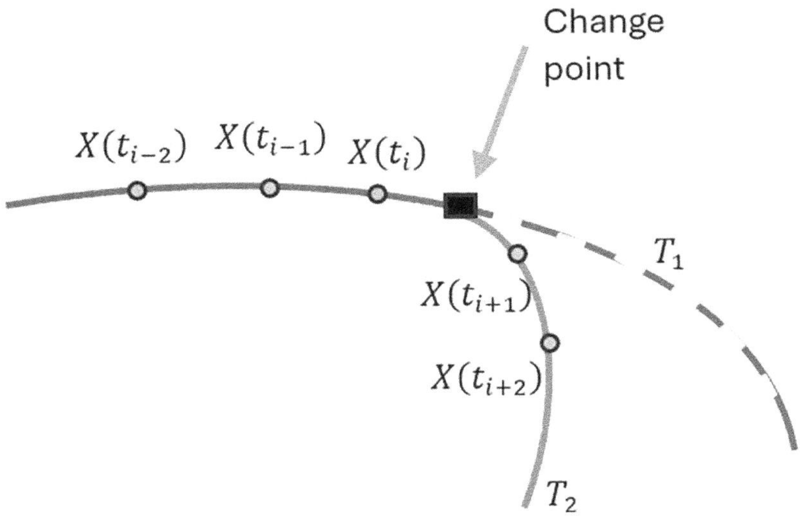

FIGURE 8.6 An illustration of a change point in a solution trajectory following changes in parameter values.

To this end, the K-nearest neighbour (K-NN) classifier can be used to detect the departure from the initial class of data to the new class. First, we define a distance metric to help form a decision boundary between data from T_1 and data from T_2. Given the set of K consecutive points $\{X(t_{i-K+1}), X(t_{i-K+2}), \ldots, X(t_{i-1}), X(t_i)\}$, the distance metric measures the average separation of a new data point from T_1, relative to its separation from the initial point $X(t_{i-K+1})$, and this is given by

$$D(t_{i+1}) = \frac{1}{K}\sum_{j=1}^{K} \frac{\| X(t_{i-j}) - X(t_{i+1}) \|}{\| X(t_{i-k+1}) - X(t_{i+1}) \|}. \tag{8.22}$$

It is expected that, if $X(t_{i+1})$ belongs to T_2, a considerable jump in the value $D(t_{i+1})$ will be observed due to the increasing separation of T_2 from T_1 as time progresses. Therefore, in order to detect the change point, it is reasonable to monitor the value $CP(t_{i+1}) = |D(t_{i+1}) - D(t_i)| / |D(t_i)|$; if this value is "small", or less than a prescribed threshold, we consider $X(t_{i+1})$ to belong to T_1. Whereas, if $CP(t_{i+1})$ exceeds the threshold value, we consider $X(t_{i+1})$ to belong to T_2, and a system reset is triggered. The basic algorithm for the change point detection is provided in Table 8.2.

TABLE 8.2 Basic Algorithm for the Change Point Detection

Given $\alpha_{opt}^{(m+1)}$ computed at time t_{m+1} using algorithm in Table 8.1, compute:
$D(t_{m+1})=\frac{1}{K}\sum_{j=1}^{K}\frac{\lVert X(t_{m-j})-X(t_{m+1})\rVert}{\lVert X(t_{m-k+1})-X(t_{m+1})\rVert}$
$CP(t_{m+1})=\lvert D(t_{m+1})-D(t_m)\rvert/\lvert D(t_m)\rvert$
If $CP(t_{m+1})>threshold$
Set $P_{m+1}^{-1}=P_0^{-1}$
End

8.4 NUMERICAL EXAMPLES

The algorithm for recursive system identification with change point detection is tested using synthetic data generated from the Lorenz system and the Van der Pol system described in section 8.2.

8.4.1 Experiment 1: The Lorenz System

In this experiment, we aim to test the ability of the algorithm to detect and recover from chaos. Here a dictionary of basis functions for the second-degree polynomial is used to detect the parameters of the Lorenz system. The reference data set is obtained from the numerical integration of the system in Example 8.3 for time $t\in[0,200]$, with the parameters set as follows:

$$\alpha_{11}=-\sigma=-10,\alpha_{12}=\sigma=10,$$
$$\alpha_{21}=\rho=100,\alpha_{22}=-1,\alpha_{26}=-1,$$
$$\alpha_{33}=-\beta=-8/3,\alpha_{35}=1,$$

with initial conditions $X(0)=(x_1(0),x_2(0),x_3(0))^T=(1,0,0)^T$. Abrupt changes in the system are introduced at $t=50$ by changing the value of ρ to 28, and at $t=150$ where ρ is changed back to 100. The changes in the time series solution and the phase-portrait due to the parametric changes is shown in Figure 8.7. With this reference signal, the system dynamic is periodic for $t=0$ to $t=50$, chaotic for $t=50$ to $t=150$ and returns to being periodic after $t=150$. In this experiment, a noise of variance $\hat{\sigma}=10^{-3}$ is added to the reference signal.

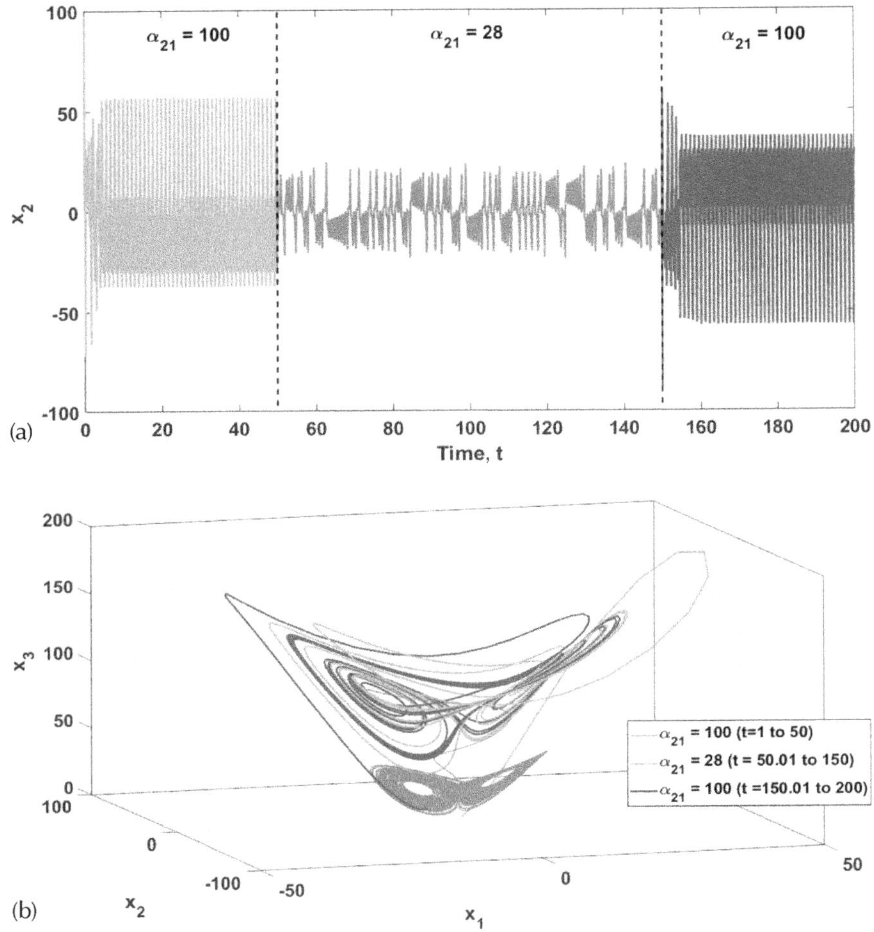

FIGURE 8.7 Reference data for Experiment 1 from the Lorenz system. (a) Time series. (b) Phase portrait.

The values of the distance metric, $D(t_i)$ produced by the algorithm in Table 8.2 are shown in Figure 8.8. It is observed that change points at $t = 50$ and $t = 150$ are clearly detected which suggests the suitability of $D(t_i)$ as a measure of class separation. The error in parameter estimation is measured using the following formula

$$\xi(t_i) = \frac{\|\alpha(t_i) - \alpha_{opt}\|_F}{\|\alpha_{opt}\|_F}, \tag{8.23}$$

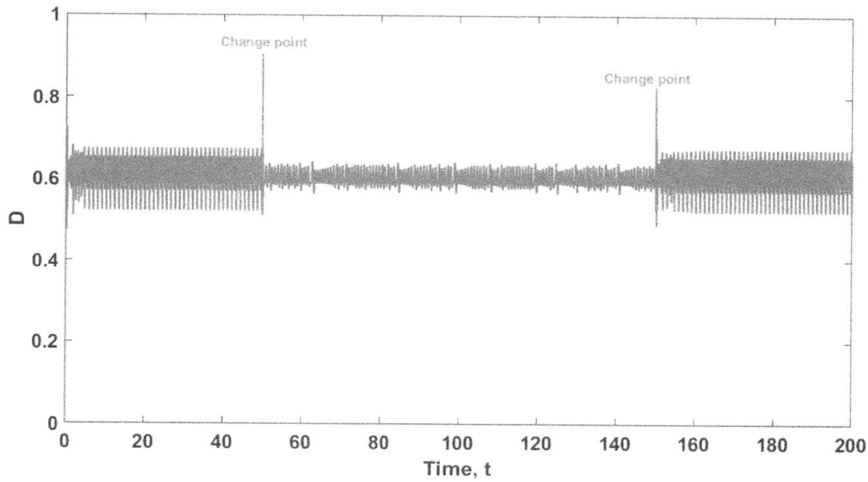

FIGURE 8.8 The distance metric for change point detection in Experiment 1.

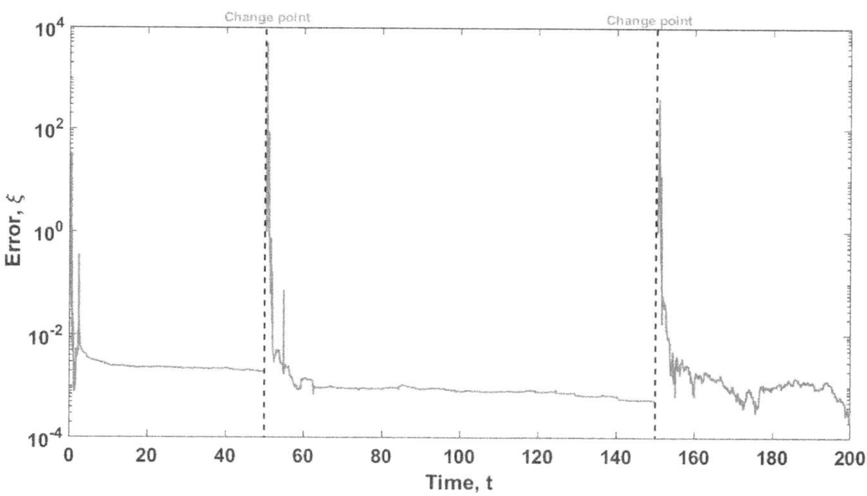

FIGURE 8.9 The error in the parameter estimation in Experiment 1.

and the values are plotted in Figure 8.9. It is observed that the minimum ξ achieved is comparable to the variance of noise in the data. The speed of recovery after abrupt parameter changes at $t = 50$ and $t = 150$ are also seen to be comparable.

In Table 8.3, the steady-state parameter values at three different times (before and after abrupt changes) are shown. These values demonstrate

TABLE 8.3 Steady-State Parameter Values of the Lorenz System Before and After Abrupt Changes

	Steady-State at $t=50$		Steady-State at $t=150$		Steady-State at $t=200$	
Parameters	**True**	**Approx.**	**True**	**Approx.**	**True**	**Approx.**
$\alpha_{11}=-\sigma$	−10	−10.0000	−10	−10.0001	−10	−10.0011
$\alpha_{12}=\sigma$	10	10.0003	10	9.9998	10	10.0011
$\alpha_{21}=\rho$	100	99.9890	28	27.9994	100	100.0038
α_{22}	−1	−0.9957	−1	−0.9995	−1	−1.0012
α_{26}	−1	−0.9998	−1	−0.9999	−1	−1.0001
$\alpha_{33}=-\beta$	−8/3	−2.6697	−8/3	−2.6662	−8/3	−2.6661
α_{35}	1	1.0000	1	0.9998	1	0.9999

the capability of the recursive system identification to identify the true parameter values with accuracies of up to two decimal places.

8.4.2 Experiment 2: Van de Pol System

In this experiment, we study the capability of the algorithm to detect structural changes in the system. A dictionary of basis functions for the third-degree polynomial is used to recover the parameters of the Van der Pol system. To induce structural changes in the system, two nonlinear terms are included in the model in Example 8.4, namely, we add a term involving $\phi_5 = x_2^2$ to the first equation, and a term involving $\phi_3 = x_1^2$ to the second equation. The modified system is given by

$$\frac{dx_1}{dt} = \alpha_{12}\phi_2 + \alpha_{15}\phi_5, \tag{8.24}$$

$$\frac{dx_2}{dt} = \alpha_{21}\phi_1 + \alpha_{22}\phi_2 + \alpha_{23}\phi_3 + \alpha_{27}\phi_7. \tag{8.25}$$

A numerical integration of the system above is performed using MATLAB to generate a reference data set for time $t \in [0,150]$, with the parameters set as follows: $\alpha_{12} = 1, \alpha_{15} = \tau, \alpha_{21} = -1, \alpha_{22} = \mu, \alpha_{23} = -\tau, \alpha_{27} = -\mu$, with initial conditions $\mathbf{x}(0) = (x_1(0), x_2(0), x_3(0))^T = (1,0,0)^T$. The system is initialised with $\tau = 0, \mu = 0.1$ where the Van der Pol system has a periodic solution. Abrupt structural changes in the system is introduced at $t = 50$ by setting $\tau = 0.14, \mu = 2.0$. For these parameter values, the system

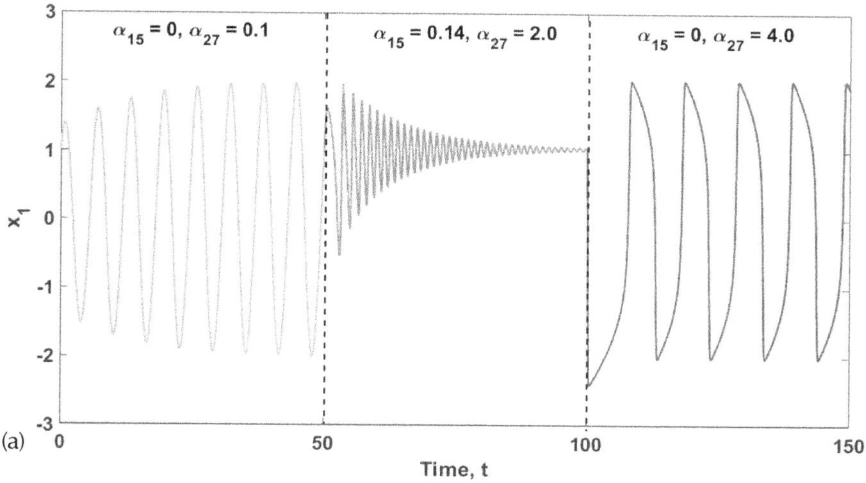

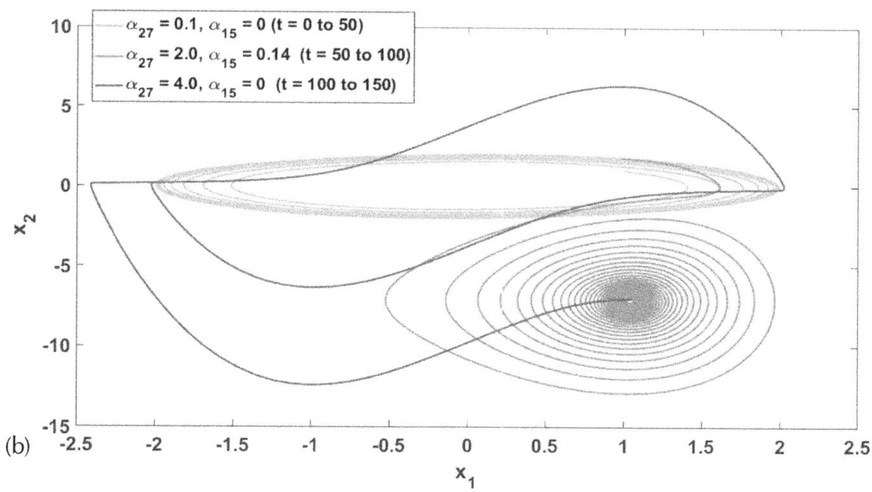

FIGURE 8.10 Reference data in Experiment 2 from the modified Van der Pol system. (a) Time series. (b) Phase portrait.

exhibits an exponentially decaying oscillatory solution. At $t = 100$, the system is returned to a periodic state by setting $\tau = 0, \mu = 4.0$. The changes in the time series solution and the phase portrait due to the parametric changes described above are shown in Figure 8.10. In this experiment, a noise of variance $\hat{\sigma} = 10^{-4}$ is added to the reference signal.

The values of the distance metric $D(t_i)$ for Experiment 2 are shown in Figure 8.11 where change points at $t = 50$ and $t = 100$ are tracked

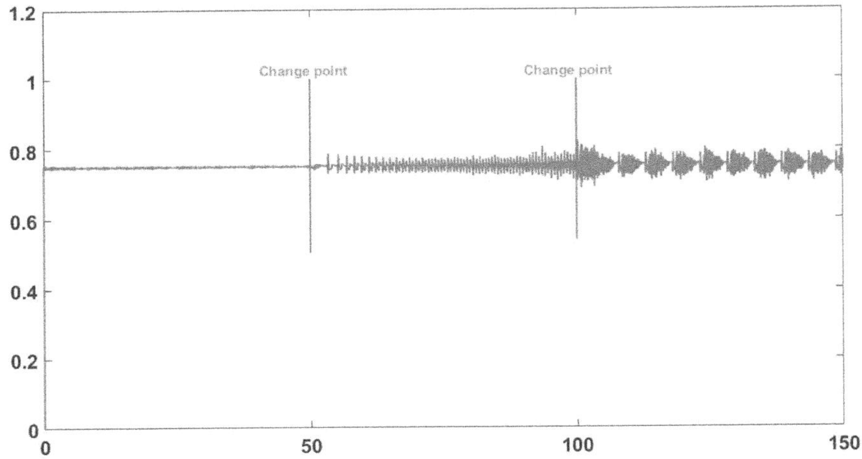

FIGURE 8.11 The distance metric for change point detection in Experiment 2.

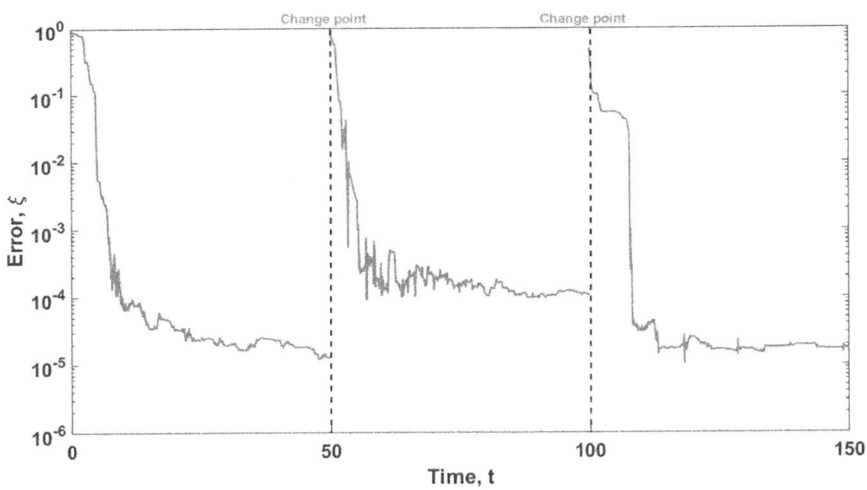

FIGURE 8.12 The error in the parameter estimation in Experiment 2.

successfully. The plot of errors ξ in Figure 8.12 shows that the minimum ξ achieved is comparable to the variance of noise in the data. The speed of recovery after abrupt structural changes at $t = 50$ and $t = 100$ are also observed to be comparable.

In Table 8.4, the steady-state parameter values at three different times (before and after abrupt changes) are shown. Accurate tracking of the true parameter values is demonstrated in this case also.

TABLE 8.4 Steady-State Parameter Values of the Van der Pol System Before and After Abrupt Changes

	Steady-State at $t = 50$		Steady-State at $t = 100$		Steady-State at $t = 150$	
Parameters	**True**	**Approx.**	**True**	**Approx.**	**True**	**Approx.**
α_{12}	1	1.0000	1	1.0000	1	1.0000
$\alpha_{15} = \tau$	0	0.0000	0.14	0.1400	0	0.0000
α_{21}	−1	−1.0000	−1	−0.9967	−1	−1.0001
$\alpha_{22} = \mu$	0.1	0.1000	2	1.9993	4	3.9999
$\alpha_{23} = -\tau$	0	0.0000	−0.14	−0.1418	0	0.0000
α_{27}	−0.1	−0.1000	−2	−2.0002	−4	−4.0000

8.5 DISCUSSION AND CONCLUSION

The mathematical formulation and algorithmic procedures of recursive nonlinear system identification with change point detection are described in this chapter. The method tracks abrupt changes in data, enabling it to respond to dynamic shifts and swiftly recover new system's parameters. The approach relies on an assumed dictionary of nonlinear basis functions that can potentially describe the dynamics of the system. The efficacy of the method is demonstrated using reference data generated from the Lorenz system and the modified Van der Pol system. Data generated from the Lorenz system is used to test the ability of the method to track the transition from periodic to chaotic dynamics, and vice versa. In the experiment involving the Van der Pol system, additional nonlinear terms are added to the system to test the capability of the algorithm to detect structural changes. In both experiments, it is shown that the change point detection mechanism implemented into the recursive system identification is effective to detect the change in the solution trajectory, bring about a reset of the recursive identification system and consequently initiate the recovery of the new system parameters that describe the new dynamic.

Nonetheless, there are limitations in the proposed recursive nonlinear system identification strategy that can be improved to better address the difficulties in handling real-life data. These include:

1. *Recursive sparse system identification*: For most systems, the parameter matrix α is sparse. The recursive identification can be made more efficient if the sparse elements of the matrix can be identified and

eliminated at each recursive step. This will be an extension of the sparse-regression approach proposed by Brunton et al. (2016) and similar in principle to the method proposed by Quade et al. (2018).

2. *Inverse-free optimisation*: The recursive identification algorithm requires the estimation of P_m^{-1} through the Sherman-Morrison formula. This approach may not be sufficiently robust to handle real-life data. An alternative method would be to adopt inverse-free incremental learning algorithm; for example, see Zhu and Wu (2020).

3. *Approximated gradient*: In the event that data for the true gradient is unavailable, approximation of the gradient using finite-difference method or total-variation regularised derivative proposed by Chartrand (2011) can be used to form matrix Y_m.

In conclusion, the applicability of the proposed framework for recursive nonlinear system identification with change point detection is clearly promising. It is possible to extend this framework to a control-oriented strategy as described in Ornik et al. (2020), where simultaneous identification of control parameters can be included in the form of an optimal control problem. With this extension, a wider range of applicability can be explored further.

8.6 ACKNOWLEDGEMENT

This work was supported by the Ministry of Higher Education Malaysia Fundamental Research Grant Scheme with Project Code: FRGS/1/2020/ICT06/USM/02/1.

REFERENCES

Brunton, S. L., Proctor, J. L., & Kutz, J. N. (2016). Discovering governing equations from data by sparse identification of nonlinear dynamical systems. *Proceedings of the National Academy of Sciences of the United States of America*, *113*(15), 3932–3937. https://doi.org/10.1073/pnas.1517384113

Chartrand, R. (2011). Numerical differentiation of noisy, nonsmooth data. *International Scholarly Research Notices*, Article ID 164564. https://doi.org/10.5402/2011/164564

Ornik, M., Carr, S., Israel, A., & Topcu, U. (2020). Control-oriented learning on the fly. *IEEE Transactions on Automatic Control*, *65*(11), 4800–4807. https://doi.org/10.1109/TAC.2019.2963293.

Quade, M., Abel, M., Kutz, J. N., & Brunton, S. L. (2018). Sparse identification of nonlinear dynamics for rapid model recovery. *Chaos*, *28*(6), 063116.

Sherman, J., & Morrison, W. J. (1950). Adjustment of an inverse matrix corresponding to a change in one element of a given matrix. *The Annals of Mathematical Statistics*, *21*(1), 124–127.

Zhang, H., Rowley, C. W., Deem, E. A., & Cattafesta, L. N. (2019). Online dynamic mode decomposition for time-varying systems. *SIAM Journal on Applied Dynamical Systems*, *18*(3), 1586–1609.

Zhu, H., & Wu, Y. (2020). Inverse-free incremental learning algorithms with reduced complexity for regularized extreme learning machine. *IEEE Access*, *8*, 177318–177328. https://doi.org/10.1109/ACCESS.2020.3022174

CHAPTER 9

A Comparative Analysis of Random Forest and ARIMA-GARCH Models

A Case Study in Forecasting Particulate Matter Concentration in Penang, Malaysia

Nuralia Nisa Mohd Nor Hisham,
Norhashidah Awang and Zainudin Arsad

9.1 INTRODUCTION

Excessive amounts of particles and harmful gases in the air can seriously harm human health, plant life, our overall well-being and even the environment, causing what we commonly refer to as air pollution. Particulate matter (PM) or particle pollution is a mixture of liquid droplets and solid particles discovered in the air. Particulate matter encompasses PM_{10} and $PM_{2.5}$, with PM_{10} denoting inhalable particles possessing diameters of 10 µm or less and $PM_{2.5}$ representing fine inhalable particles with diameters measuring 2.5 µm or less. Particulate matter comprises microscopic liquid droplets or solids which can be inhaled and can affect health in a serious manner (United States Environmental Protection Agency, 2022).

DOI: 10.1201/9781032678573-9

Therefore, it is necessary to estimate the future concentration of particulate matter, and the best statistical technique must be investigated for obtaining the most accurate prediction of the particulate matter concentration. The assessment and improvement of air quality is of paramount importance in various engineering applications, where precise environmental monitoring and control measures are essential for ensuring the health and safety of both humans and infrastructure.

9.2 BACKGROUND INFORMATION

Many methods have been proposed by researchers and environmental agencies for air quality analysis, but there is no universally accepted method that can properly explain all situations (Sharma et al., 2021). Therefore a comparative study of techniques needs to be conducted to find the best technique which is able to predict the particulate matter concentration in Penang, Malaysia, from July 2017 to December 2020 with the best performance evaluation measures. Within the realm of prior research endeavors concerning air pollutant data, several studies have employed the random forest (RF), autoregressive moving average (ARMA), autoregressive integrated moving average (ARIMA) and ARIMA–generalised autoregressive conditional heteroskedasticity (ARIMA-GARCH) techniques for the purpose of modeling and forecasting forthcoming air pollutant values.

Kumar and Jain (2010) forecasted daily mean ambient air pollutants concentration, which are O_3, CO, NO and NO_2, at an urban traffic site (ITO) in Delhi, India, by applying a stationary stochastic ARIMA modelling approach. Masseran (2017) conducted an analysis of hourly PM_{10} data within the context of Kuala Lumpur, Malaysia. In this chapter, an ARMA-GARCH framework is employed to derive an adequate model capable of mitigating the impact of volatility inherent in the dataset. The ARMA(1,0) model was decided as a good estimation of the PM_{10} data, but the presence of the volatility effect has decreased the accuracy of the model. Consequently, the GARCH model was integrated into the ARMA model to address the issue of volatility. The statistical properties produced by simulated data of the ARMA(1,0)–GARCH(1,1) model are found to be similar to the original data. The inclusion of the GARCH model, rather than relying solely on an ARMA model, proved to enhance the forecasting performance of PM_{10}.

Beside the statistical methods, various studies have been conducted by applying the machine learning techniques to analyse air quality and environmental data. Zaman et al. (2021) conducted a study utilising both RF and support vector regression (SVR) to ascertain the concentrations of $PM_{2.5}$ across Malaysia during the period spanning 2018 to 2019. The estimation of $PM_{2.5}$ concentrations is based on meteorological parameters (relative humidity, air temperature, wind speed and direction), ground measured air pollutants (SO_2, O_3, NO_2, CO) and satellite AOD (aerosol optical depth) observations. The estimated $PM_{2.5}$ concentrations in the study by Zaman et al. (2021) were compared with measurements conducted at 65 air-quality monitoring stations situated at rural, suburban, urban and industrial sites. The $PM_{2.5}$ concentrations were predicted by seven models separately for suburban/rural sites and urban/industrial sites, for the four dominant season (two inter-monsoon, wet and dry) and an overall model, with accuracy of R^2 between 0.46 and 0.76. The findings revealed that the RF model with accuracy of R^2 between 0.53 and 0.76 has slightly better performance than SVR. In another study, Ivanov et al. (2018) explored the application of the RF method in modeling PM_{10} concentrations in Blagoevgrad, Bulgaria. The dataset involved nine years of average daily data from 2009 to 2018, encompassing PM_{10} concentrations and six meteorological variables. The achieved normalised coefficient of determination stands at 93%, indicating a highly favorable model fit.

The RF model boasts several advantages, including its simplicity, ease of parallelisation and high accuracy. Additionally, it exhibits relative robustness to noise and outliers while also offering valuable internal assessments of variable importance, strength, correlation and error (Breiman, 2001).

Drawing from the preceding studies, the approach used in this chapter compares two distinct techniques from two different paradigms: statistical and machine learning. The statistical approach that is adopted here is the univariate ARIMA-GARCH, while for the machine learning approach, RF is chosen. These methods are employed to analyse PM_{10} and $PM_{2.5}$ concentrations at four air monitoring stations located in both urban and suburban areas within Penang, Malaysia. Following that, a comparative analysis of these two techniques is conducted to determine which one yields the more accurate forecasts for PM_{10} and $PM_{2.5}$ concentrations.

9.3 METHODS

This section explains the methods to be used in this study, which are ARIMA-GARCH and RF techniques. In order to understand the RF method, the classification and regression tree (CART) is explained first, as the RF method is the extended method from CART decision trees.

9.3.1 CART for Creating a Regression Tree

The RF method, developed by Leo Breiman in 2001, is a notable ensemble learning technique that generates an ensemble of tree-structured classifiers. This method extends and combines the capabilities of CART decision trees. It offers a robust and powerful approach to classification and regression tasks, contributing to the field of machine learning and data analysis.

CART is a powerful analytical tool capable of handling both categorical data, leading to the creation of classification trees, and continuous data, resulting in regression trees. It operates as a binary tree, iteratively dividing nodes into two child nodes starting from the root, encompassing the entire learning sample (Roy et al., 2016). For terminal nodes, where no further splits occur, numerical values can be assigned using either the sample mean or sample median of the response values associated with the members of the learning sample in that node (Sutton, 2005). The split chosen at each stage of tree-growing process is based on the greatest reduction in the sum of the squared differences between the response values for the learning sample cases corresponding to a particular node and their sample mean. For instance, squared differences criterion is used to find the plane which divides a subset of X (root node) into the sets A (right node) and B (left node) for which the function below is minimised:

$$\sum_{i:x_i \in A}\left(y_i - \bar{y}_A\right)^2 + \sum_{i:x_i \in B}\left(y_i - \bar{y}_B\right)^2, \tag{9.1}$$

where $\bar{y}_A$ represents the sample mean of the response values for cases in the learning sample corresponding to A and $\bar{y}_B$ represents the sample mean of the response values for cases in the learning sample corresponding to B. Pruning process is needed to reduce overfitting of a decision tree as pruning removes the weakest rules to simplify the decision tree (Krueger, 2021).

9.3.2 Random Forest

In RF, the growth of each tree in the ensemble involves the use of randomly selected inputs or combinations of inputs at each node. RFs employ a combination of bagging and random feature selection. Bagging entails the creation of a new training set drawn from the original training set with replacement. In RF, the trees are expanded using the CART methodology, allowing them to grow to their maximum size without pruning (Breiman, 2001). For example, a bootstrap training set T_k, where k is number of trees, is formed from a particular training set T. Classifiers $h(\mathbf{x}, T_k)$, where $\mathbf{x}$ is an input vector, are constructed, and the bagged predictor is formed from their votes. The votes are aggregated only for those classifiers where T_k does not contain $(y, \mathbf{x})$, where y is the output or response variable, for each $(y, \mathbf{x})$ in the training set. These are called *out-of-bag classifiers*. The error rate of the out-of-bag classifiers on the training set is the out-of-bag estimate for the generalisation error (Breiman, 2001). *Out-of-bag* (OOB) can be understood as the data points which fail to be selected as training set during the bagging process (Radhika, 2020).

The contribution of each predictor variable (feature) in response variable estimations is analysed by considering variable importance statistic. There are two popular measures of feature importance for the RF model which are mean decrease accuracy (MDA) and mean decrease Gini (MDG). MDA is calculated based on OOB samples, while MDG is calculated based on Gini importance which the average gain by splits of a given variable is measured (Zaman et al., 2021). The scikit-learn package in Python can be used to build the RF model, but it is important to bear in mind that scikit-learn does not offer the capability to implement the Mean Decrease Accuracy (MDA) measure, so to assess feature importance, the MDG metric can be used instead. Within the scikit-learn package, the importance of nodes for each decision tree is calculated using the Gini Importance method.

9.3.3 Univariate ARIMA-GARCH

Univariate ARMA is a popular model used to analyse time series data in many fields of research, including air pollution, economics and medicine. ARMA modelling comprises three iterative steps to obtain an optimal model: identification, estimation and diagnostic checking (Li et al., 2020). This method assumes the time series to be analysed are stationary. The ARIMA can be viewed as an extension of ARMA models, where differencing is used to transform a non-stationary time series into a stationary one, prior to applying ARMA techniques. Usually, it is sufficient to apply a low order of differencing, d, which is 1 or 2 at most, to produce a time series that is stationary in the mean. The estimated autocorrelation functions (ACF) of the original series and the differenced series need to be examined. The estimated ACF of a non-stationary series normally tails off extremely slowly in linear fashion, which means the ACF of the original non-stationary series is usually highly significant even at high lags.

The primary objective during the identification stage is to match patterns in the sample ACF and partial ACF (PACF) with patterns associated with a few known underlying processes. Theoretically, if the ACF tails off towards zero (exponential, alternatingly or sine wave) and PACF cuts off at lag p, then the AR(p) model is estimated for the series. Meanwhile, MA(q) model is estimated if the ACF cuts off at lag q and the PACF tails off to zero. Tailing off in both the ACF and PACF indicate that ARMA(p,q) is an appropriate model to be estimated. ARIMA(p,d,q) in the general form can be defined as follows:

$$\varphi(B)(1-B)^d Y_t = \theta(B)\varepsilon_t, \tag{9.2}$$

where Y_t is the observed value at time t, ε_t is the random error term at time t, $\varphi_1, \varphi_2, \ldots, \varphi_p$ are the autoregressive coefficients of the pth degree polynomial $\varphi(B) = 1 - \varphi_1 B - \varphi_2 B^2 - \ldots - \varphi_p B^p$, $\theta_1, \theta_2, \ldots, \theta_p$ are the moving average coefficients of the qth degree polynomial $\theta(B) = 1 - \theta_1 B - \theta_2 B^2 - \ldots -\theta_p B^q$, d is the order of differencing and B is the backward shift operator (Hussin et al., 2021).

Estimation of the ARIMA model is typically performed using the maximum likelihood method. As part of the model evaluation process, testing the goodness-of-fit is conducted for model diagnostics, and appropriate modifications are suggested in cases where the fit is poor. This iterative process aims to develop a statistically superior model that can accurately capture the underlying series dynamics. A fitted ARIMA model is considered statistically adequate when the parameter estimates are believed to closely approximate the true values. This is often indicated by residuals possessing the characteristics of a white noise process, which includes independence, identical distribution, normality, zero mean and constant variance. Normality of the residuals can be determined by conducting a normality test, such as the Anderson-Darling test. Next, a test that takes into consideration the magnitude of residual autocorrelations as a group is useful to be considered. A sample ACF of the residuals as a unit or group is used by the Portmanteau and Ljung-Box-Pierce (LB) lack-of-fit tests to determine the following joint null hypothesis (Wei, 2006):

$$H_0 : \rho_{1,\varepsilon} = \rho_{2,\varepsilon} = \cdots = \rho_{K,\varepsilon} = 0$$

where $\rho_{i,\varepsilon}$ is the residual autocorrelations magnitude at lag i.

The residuals from an ARIMA model can be tested for time-varying variation (heteroscedastic) by a test that is based on the Lagrange multiplier, better known as the ARCH-LM test. Heteroscedasticity may be solved by estimating the GARCH model for the variation of the series. When there are "competing candidates" among the estimated models, model selection criteria is often used to determine the most appropriate model. One of the popular measures for trading off the goodness-of-fit against extra number of parameters is the Schwartz Bayesian Criterion (SBC). A better model among the models constructed with different lags can be identified by lower SBC values (Li et al., 2020). The error process ε_t, satisfying the AR(m), is described as an autoregressive conditionally heteroscedastic process of order m, known as ARCH(m). The concept of ARCH was initially introduced by economist Engle (1982). The ARCH(m) process can be generalised to a process in which the conditional variance depends on an infinite number of lags of ε_{t-j}^2, giving rise to the generalised ARCH (GARCH) process, which was developed by Bollerslev (1986). The GARCH of order r and s, GARCH(r,s) can also be interpreted as an ARMA(r,s) process for the conditional variance. For example, the equation of ARMA(1,1)-GARCH(1,1) is written as

$$Y_t = \phi_1 Y_{t-1} + \varepsilon_t - \theta_1 \varepsilon_{t-1},\ \varepsilon_t \sim \sigma_t \upsilon_t \text{ such that } \upsilon_t \sim NID(0,1) \quad (9.3)$$

and

$$\sigma_t^2 = \alpha_0 + \alpha_1 \varepsilon_{t-1}^2 + \beta_1 \sigma_{t-1}^2. \quad (9.4)$$

9.4 RESULTS AND DISCUSSION

The RF method and the univariate ARIMA-GARCH method are used to model the daily particulate matter concentrations in Penang, Malaysia. Penang is chosen as the study location due to its status as one of the most rapidly developing states characterised by a proliferation of industrial activities, significant urban development and a densely populated environment. Consequently, an elevated level of air pollutant concentrations within the region is anticipated. Four air monitoring stations have been strategically selected to represent both urban and suburban areas within the region. These stations are Sekolah Kebangsaan Seberang Jaya II in Seberang Jaya, Kolej Vokasional Seberang Perai in Seberang Perai, Universiti Sains Malaysia in Minden and Kolej Vokasional Balik Pulau in

TABLE 9.1 The Start Date, End Date and Number of Observations for All Stations

Station	Start Date	End Date	Number of Observations
Seberang Jaya	24 July 2017	31 December 2020	1257
Seberang Perai	25 July 2017	31 December 2020	1256
Minden	1 March 2018	31 December 2020	1037
Balik Pulau	26 July 2017	31 December 2020	1255

Balik Pulau; they serve as vital data collection points. It is noteworthy that Seberang Jaya and Minden are classified as urban areas, while Seberang Perai and Balik Pulau fall within the suburban category.

The dataset for analysis consists of 12 daily average measured variables sourced from the Department of Environment Malaysia. These variables include concentrations of PM_{10}, $PM_{2.5}$, sulfur dioxide (SO_2), nitrogen dioxide (NO_2), nitrogen monoxide (NO), nitrogen oxides (NO_x), ozone (O_3) and carbon monoxide (CO), as well as measurements of relative humidity, ambient temperature, wind speed and wind direction.

It is important to note that the study period varies among the monitoring stations, and this temporal diversity is designed to prevent data gaps. The data collection at the four monitoring stations is summarised in Table 9.1.

In this study, the RF algorithm is applied using Python in the Jupyter Notebook environment. On the other hand, the univariate ARIMA-GARCH analysis is conducted using multiple software tools, including EViews, SPSS and Minitab. EViews is primarily utilised for the majority of the univariate ARIMA-GARCH analysis. Minitab plays a crucial role in performing data descriptive tasks, while SPSS is instrumental in constructing correlogram graphs, plotting residuals and aiding in the detection of outliers.

In line with the approach employed by Ivanov et al. (2018), linear interpolation is used to fill in missing values. The last 15 observations, spanning 17–31 December 2020, are used to compute forecasting error. The forecast performance of all models at each station is evaluated using three measures: root mean square error (RMSE), mean absolute error (MAE) and mean absolute percentage error (MAPE). A model with smaller or close-to-zero values of RMSE, MAE and MAPE is considered to be a superior model.

9.4.1 RF Analysis

In this study, two new variables, namely the one-day ahead concentration of PM_{10} (PM_{10} next day) and the one-day ahead concentration of $PM_{2.5}$

($PM_{2.5}$ next day), have been created. These two new variables have been designated as the dependent variables, while the remaining 12 variables are selected as the independent variables. As a result, two separate models with different dependent variables have been generated for each station. This study examines how the current measurements of independent variables influence the measurement of the dependent variables on the following day.

In this analysis, the number of regression trees is limited to 1000, a choice influenced by the work of Ivanov et al. (2018). Furthermore, a comparison is conducted for both the dataset with outliers and the dataset with adjusted outliers concerning the dependent variables, namely PM_{10} and $PM_{2.5}$. The identification of outliers and extreme values has been accomplished by examining the boxplot. While a substantial number of outliers are detected, to prevent a significant loss of information, only the extreme values are addressed. These extreme values are handled using linear interpolation, following the recommendation by Wahir et al. (2018). Consequently, two distinct models are established: "W Extreme" for the dataset using extreme values and "Adj Extreme" for the dataset incorporating data with the extreme values adjusted via linear interpolation. The summary of the RF analysis is shown in Table 9.2.

The prediction results in Table 9.2 show that there is no discernible difference in the RMSE, MAPE and MAE values between the W Extreme data and Adj Extreme data for both PM_{10} and $PM_{2.5}$ models across all stations. Consequently, there is no need to address extreme values in order to achieve an improved predictive model using the RF technique.

Furthermore, it is of paramount importance to assess the strength of the correlation between one-day ahead concentrations of PM_{10} and $PM_{2.5}$ and all the variables employed in their prediction. This analysis can indirectly reveal insights into the pollution process and its sources. The significance of the explanatory variables, or their feature importance, is evaluated using methods such as Gini importance or mean decrease impurity. Bar graphs depicting the Gini importance scores for each feature across the four categories (Adj Extreme PM_{10}, Adj Extreme $PM_{2.5}$, W Extreme PM_{10} and W Extreme $PM_{2.5}$) are generated and presented in Figures 9.1 to 9.4.

For the Seberang Jaya station (Figure 9.1), it is generally observed that PM_{10} and $PM_{2.5}$ emerge as the two most crucial features when constructing the one-day ahead prediction models for both PM_{10} and $PM_{2.5}$, respectively.

In the case of Seberang Perai (Figure 9.2), PM_{10} is the most influential feature when forecasting one-day ahead of PM_{10} for both models.

TABLE 9.2 Summary of the Random Forest Analysis

		Test	Prediction		
Station	**Category**	**RMSE**	**RMSE**	**MAPE**	**MAE**
Seberang Jaya	W Extreme PM_{10}	7.060	4.688	16.810	3.478
	Adj Extreme PM_{10}	6.465	4.896	17.106	3.550
	W Extreme $PM_{2.5}$	5.614	3.686	19.600	2.567
	Adj Extreme $PM_{2.5}$	5.275	3.798	20.068	2.652
Seberang Perai	W Extreme PM_{10}	7.991	4.433	27.997	3.634
	Adj Extreme PM_{10}	7.994	4.446	28.085	3.645
	W Extreme $PM_{2.5}$	6.976	3.774	30.852	2.915
	Adj Extreme $PM_{2.5}$	5.689	3.718	30.716	2.920
Minden	W Extreme PM_{10}	11.570	3.331	15.731	2.605
	Adj Extreme PM_{10}	6.385	3.356	15.888	2.628
	W Extreme $PM_{2.5}$	10.445	2.737	19.023	2.191
	Adj Extreme $PM_{2.5}$	5.128	2.728	19.089	2.181
Balik Pulau	W Extreme PM_{10}	13.108	3.694	15.652	2.892
	Adj Extreme PM_{10}	6.109	3.730	15.638	2.906
	W Extreme $PM_{2.5}$	11.814	3.356	22.309	2.482
	Adj Extreme $PM_{2.5}$	5.093	3.412	22.552	2.508

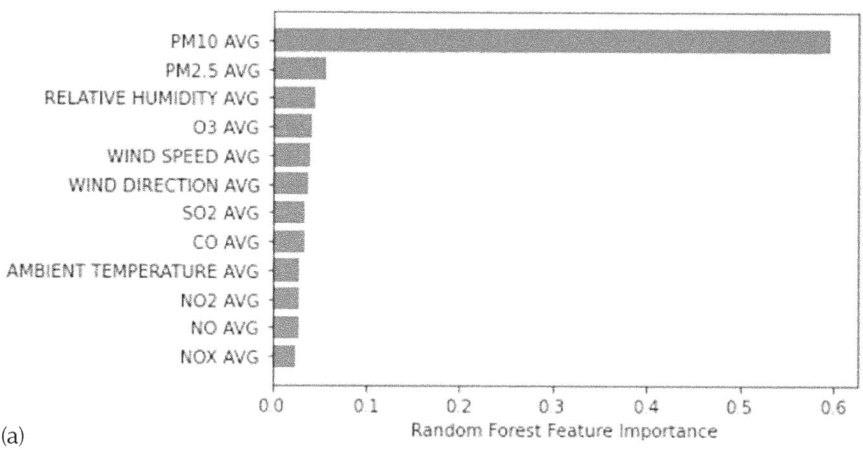

(a)

FIGURE 9.1 Graph of features importance for Seberang Jaya. (a) Adj Extreme PM_{10}. (b) Adj Extreme $PM_{2.5}$. (c) W Extreme PM_{10}. (d) W Extreme $PM_{2.5}$.

(Continued)

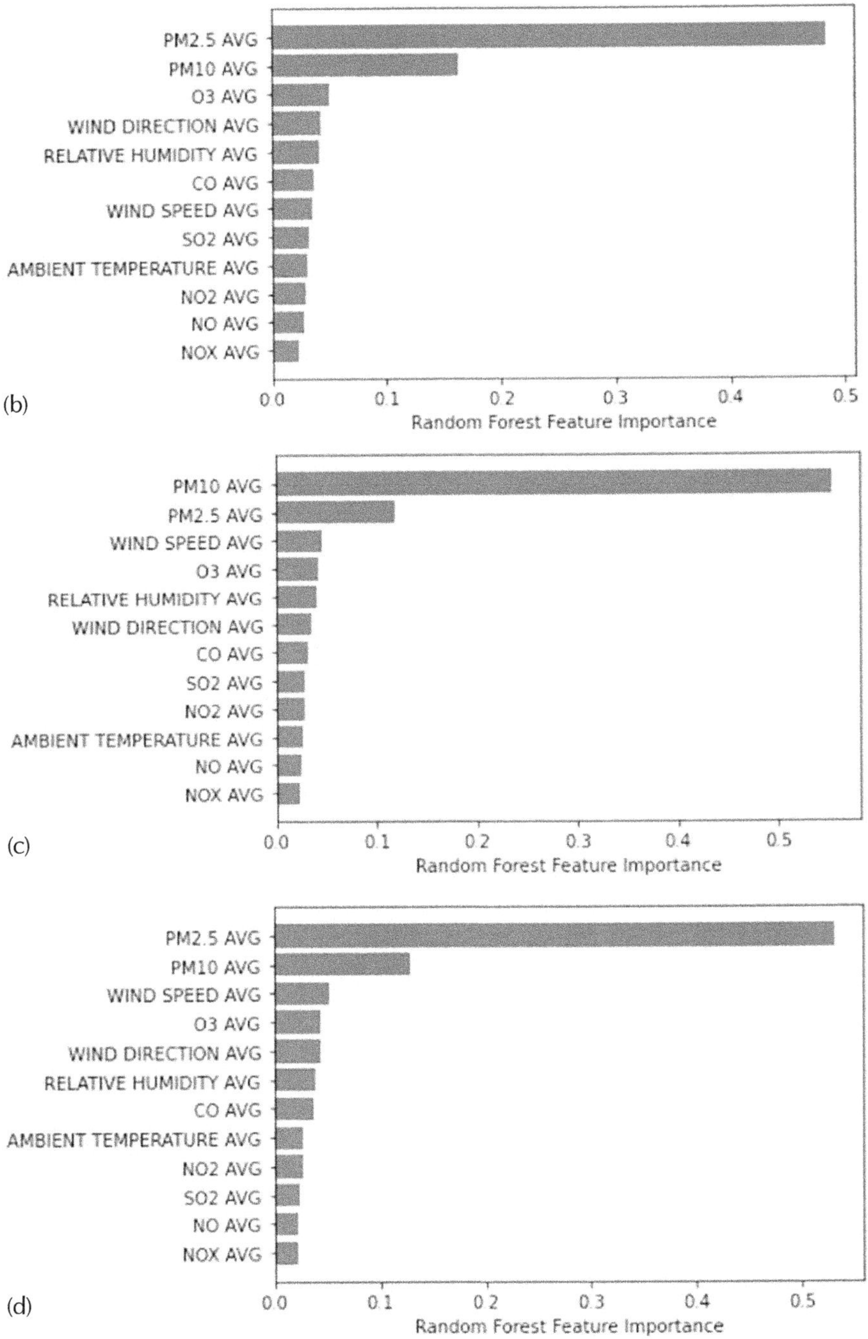

FIGURE 9.1 **(Continued)**

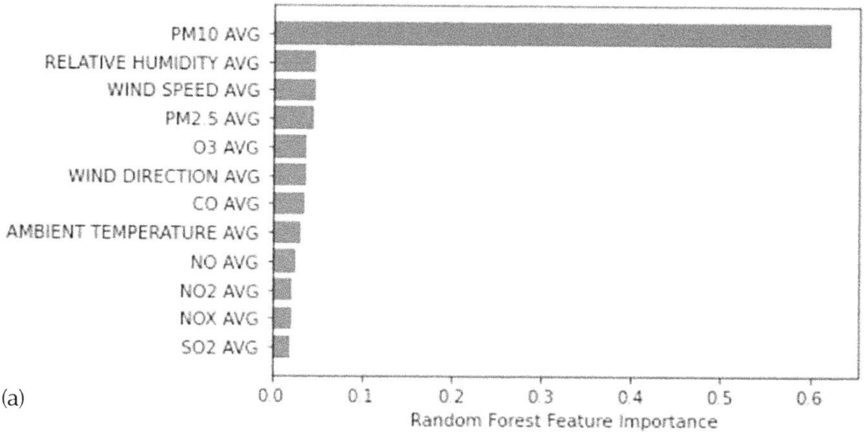

(a)

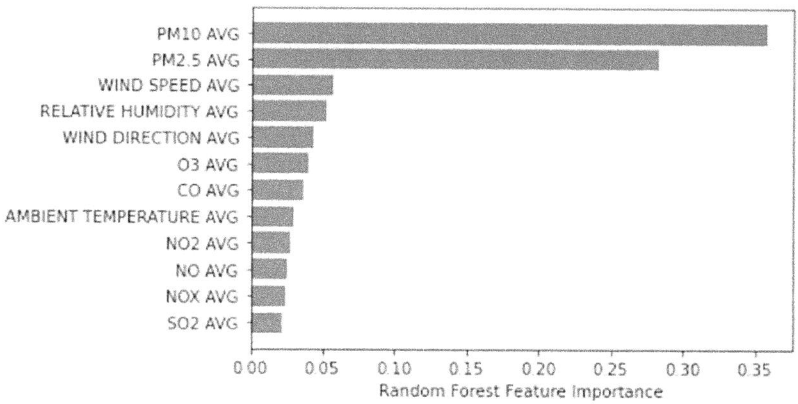

(b)

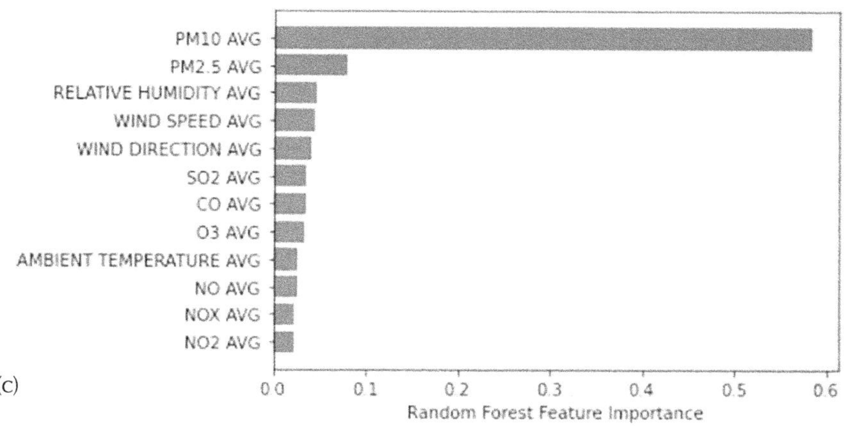

(c)

FIGURE 9.2 Graph of features importance for Seberang Perai. (a) Adj Extreme PM_{10}. (b) Adj Extreme $PM_{2.5}$. (c) W Extreme PM_{10}. (d) W Extreme $PM_{2.5}$.

(Continued)

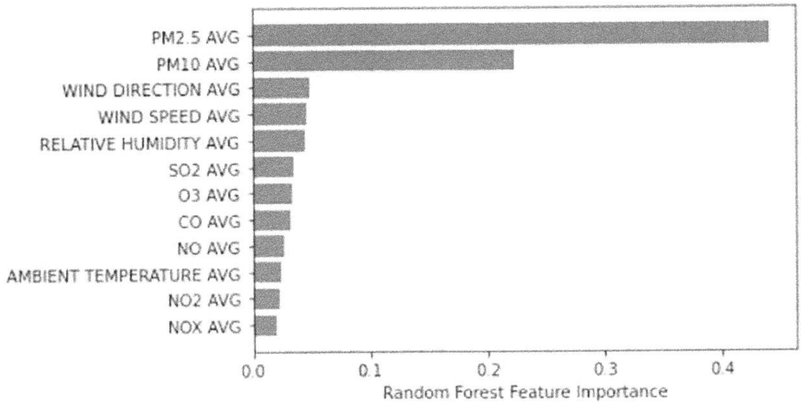

(d)

FIGURE 9.2 (Continued)

However, the second most important feature differs between the two models. Relative humidity is the second important feature for the Adj Extreme PM_{10} model, while $PM_{2.5}$ is selected as the second important feature for the W Extreme PM_{10} model. For one-day ahead prediction of $PM_{2.5}$, PM_{10} and $PM_{2.5}$ stand out as the most significant features for both the Adj Extreme and W Extreme models.

In contrast to the findings for the previously discussed stations, the Minden (Figure 9.3) and Balik Pulau (Figure 9.4) stations exhibit a different pattern. In both Minden and Balik Pulau, PM_{10} emerges as the most critical feature in both one-day ahead forecast of PM_{10} and $PM_{2.5}$ for both models. Additionally, the second most important feature for the prediction of both one-day ahead of PM_{10} and $PM_{2.5}$ is consistent in Minden, with $PM_{2.5}$ being selected. However, for Balik Pulau, CO and $PM_{2.5}$ share the position of the second most important features in predicting one-day ahead of PM_{10} and $PM_{2.5}$, respectively, for both models.

Across all stations, it is generally observed that PM_{10} and $PM_{2.5}$ consistently emerge as the two most crucial features for predicting both one-day ahead of PM_{10} and $PM_{2.5}$ levels. This analysis reveals that the treatment of extreme values primarily affects the Seberang Perai station, while for the other stations, most models indicate the same two critical features for both one-day ahead predictions of PM_{10} and $PM_{2.5}$.

9.4.2 Univariate ARIMA-GARCH Analysis

For ARIMA-GARCH modelling, natural logarithm and square root transformations are used on all time series to produce nearly normally

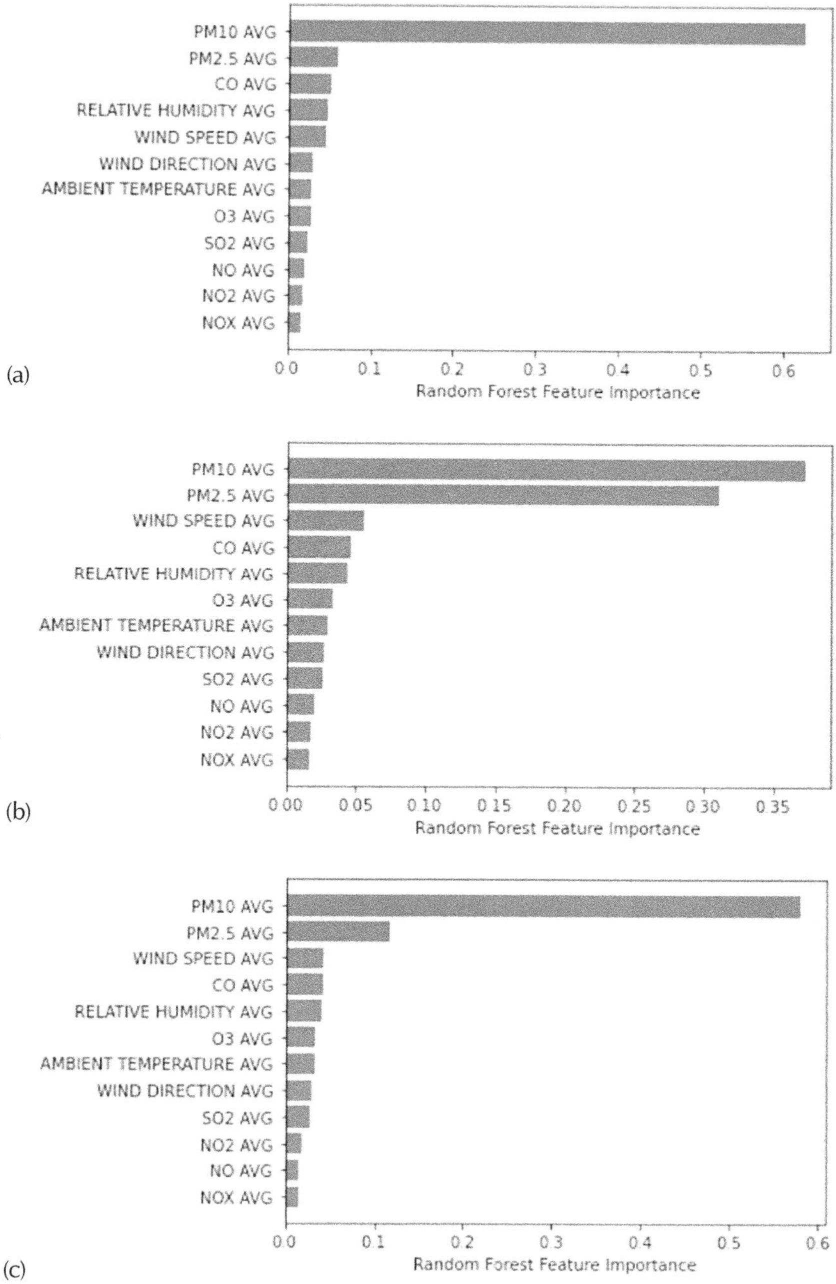

FIGURE 9.3 Graph of features importance for Minden. (a) Adj Extreme PM_{10}. (b) Adj Extreme $PM_{2.5}$. (c) W Extreme PM_{10}. (d) W Extreme $PM_{2.5}$.

(Continued)

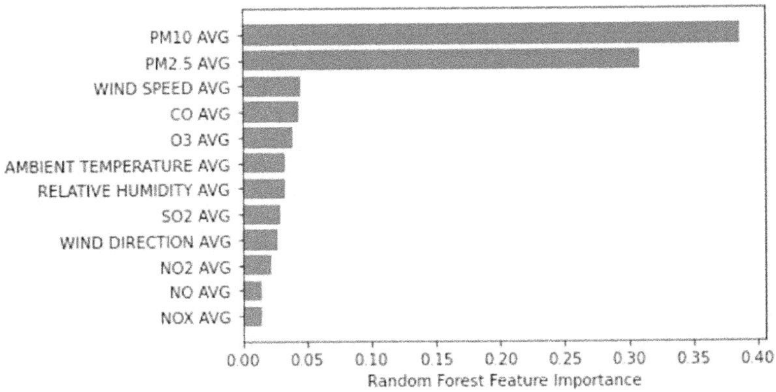

(d)

FIGURE 9.3 (Continued)

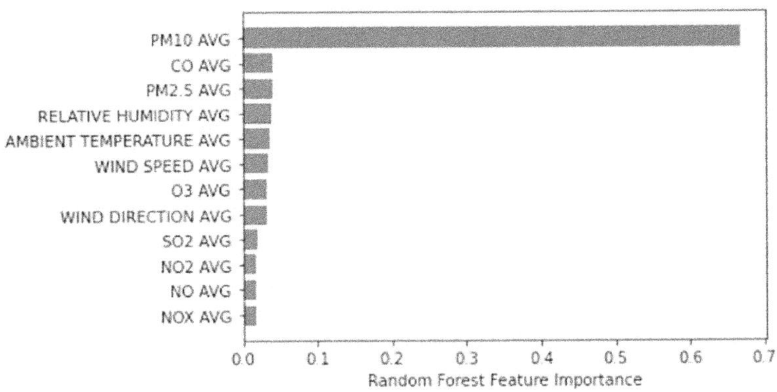

(a)

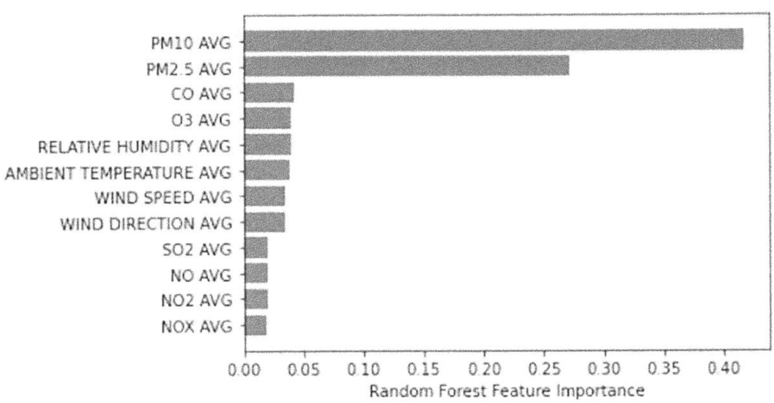

(b)

FIGURE 9.4 Graph of features importance for Balik Pulau. (a) Adj Extreme PM_{10}. (b) Adj Extreme $PM_{2.5}$. (c) W Extreme PM_{10}. (d) W Extreme $PM_{2.5}$.

(Continued)

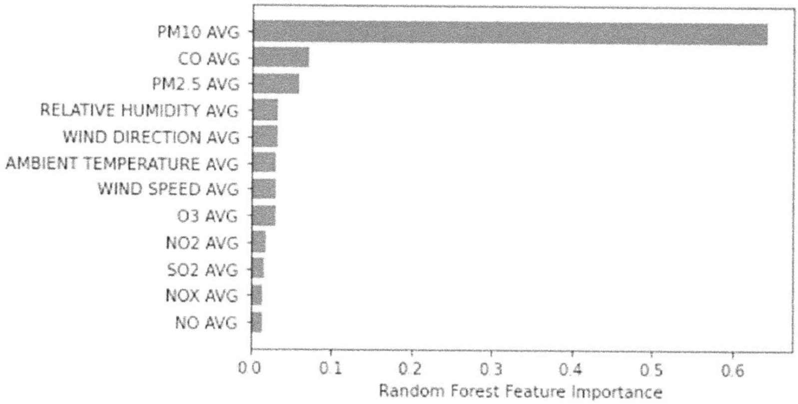

(c)

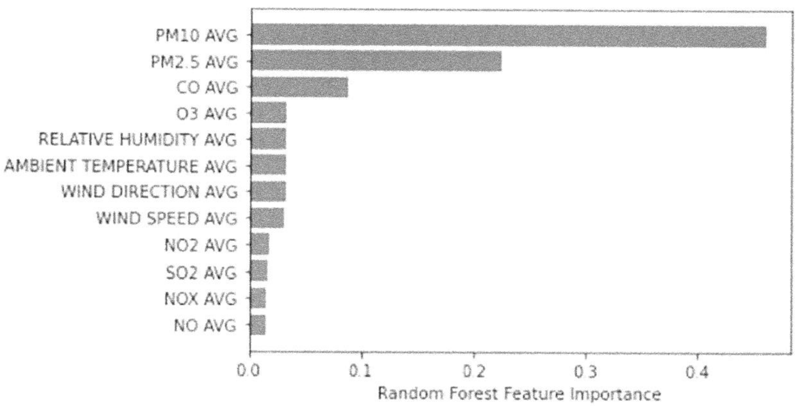

(d)

FIGURE 9.4 (Continued)

distributed data. Table 9.3 summarises the best ARMA or ARMA-GARCH model for LN PM_{10} and LN $PM_{2.5}$, where LN indicates the natural logarithm transformation of the both variables for all stations.

The results in Table 9.3 are compared with the case when outliers and extreme values are adjusted to determine whether the ARMA-GARCH model is affected by outliers and extreme values. The summary of the best model for adjusted outliers' data is shown in Table 9.4 and the summary of forecasting performance measures for all models is shown in Table 9.5.

The term "Ori" in Table 9.5 refers to the original time series, while "Adj Outlier" denotes the time series after the outlier treatment. In general, all stations, with the exception of Minden, exhibit no significant variation in

TABLE 9.3 Summary of the Best ARMA or ARMA-GARCH Model for LN PM_{10} and LM $PM_{2.5}$ for All Stations

Station	Variable	Best Model
Seberang Jaya	LN PM_{10}	ARMA(3,1)-GARCH(1,1)
	LN $PM_{2.5}$	ARMA(1,3)
Seberang Perai	LN PM_{10}	ARMA(3,1)-GARCH(1,1)
	LN $PM_{2.5}$	ARMA(2,2)-GARCH(1,1)
Minden	LN PM_{10}	ARMA(2,2)-GARCH(1,1)
	LN $PM_{2.5}$	ARMA(2,2)-GARCH(1,1)
Balik Pulau	LN PM_{10}	ARMA(3,1)
	LN $PM_{2.5}$	ARMA(3,1)

TABLE 9.4 Summary of the Best Model for Adjusted Outliers' Data

Station	Variable	Best Model
Seberang Jaya	LN PM_{10}	ARMA(2,1)
	LN $PM_{2.5}$	ARMA(2,1)-GARCH(1,1)
Seberang Perai	LN PM_{10}	ARMA(3,1)
	LN $PM_{2.5}$	ARMA(2,1)
Minden	LN PM_{10}	ARMA(3,2)
	LN $PM_{2.5}$	ARMA(2,1)-GARCH(1,1)
Balik Pulau	LN PM_{10}	ARMA(3,1)
	LN $PM_{2.5}$	ARMA(3,1)

TABLE 9.5 Summary of Forecasting Performance Evaluation Measures for All Models

			Prediction		
Station	PM	Model	RMSE	MAPE	MAE
Seberang Jaya	PM_{10}	Ori	4.7594	20.2577	4.3335
		Adj Outlier	4.8445	20.7375	4.4218
	$PM_{2.5}$	Ori	3.7795	26.9472	3.5319
		Adj Outlier	3.8699	27.5341	3.6310
Seberang Perai	PM_{10}	Ori	5.0692	17.3667	4.2109
		Adj Outlier	5.1241	17.4101	4.2356
	$PM_{2.5}$	Ori	3.7887	20.5152	3.1469
		Adj Outlier	3.9255	20.0034	3.1428
Minden	PM_{10}	Ori	6.4383	18.8204	4.8911
		Adj Outlier	6.7754	19.6803	5.1324
	$PM_{2.5}$	Ori	4.3318	18.3062	3.1548
		Adj Outlier	4.6146	19.0404	3.3250
Balik Pulau	PM_{10}	Ori	5.5945	18.0702	4.4518
		Adj Outlier	5.6880	18.0434	4.4781
	$PM_{2.5}$	Ori	4.1826	19.3137	3.1621
		Adj Outlier	4.2128	19.0140	3.1449

the values of RMSE, MAPE and MAE for the two models of $PM_{10.}$ and $PM_{2.5}$. Although there is a slight disparity in RMSE and MAE, the Ori model consistently yields smaller values of RMSE and MAE.

In summary, within the scope of this study, it can be concluded that, in terms of forecasting, employing the original series for modelling is better across all stations.

9.4.3 Comparison between RF Model and Univariate ARIMA-GARCH Model

In this section, the best result obtained from the RF model and the univariate ARMA-GARCH is compared based on their performance in forecasting. The comparison is conducted by assessing the forecasting performance using three evaluation measures, RMSE, MAPE and MAE. The summary of these evaluation measures for all models is presented in Table 9.6.

Based on the data presented in Table 9.6, it is evident that for the Minden and Balik Pulau stations, the RMSE values obtained from the ARMA-GARCH model are notably larger than those of the RF model for both PM_{10} and $PM_{2.5}$ variables. Additionally, the Seberang Perai station

TABLE 9.6 Summary of Evaluation Measures for All Models

			Prediction		
Station	**Model**	**Method**	**RMSE**	**MAPE**	**MAE**
Seberang Jaya	PM_{10}	RF	4.688	16.810	3.478
		ARMA-GARCH	4.759	20.258	4.334
	$PM_{2.5}$	RF	3.686	19.600	2.567
		ARMA-GARCH	3.780	26.947	3.532
Seberang Perai	PM_{10}	RF	4.433	27.997	3.634
		ARMA-GARCH	5.069	17.367	4.211
	$PM_{2.5}$	RF	3.774	30.852	2.915
		ARMA-GARCH	3.789	20.515	3.147
Minden	PM_{10}	RF	3.331	15.731	2.605
		ARMA-GARCH	6.438	18.820	4.891
	$PM_{2.5}$	RF	2.737	19.023	2.191
		ARMA-GARCH	4.332	18.306	3.155
Balik Pulau	PM_{10}	RF	3.694	15.652	2.892
		ARMA-GARCH	5.595	18.070	4.452
	$PM_{2.5}$	RF	3.356	22.309	2.482
		ARMA-GARCH	4.183	19.314	3.162

demonstrates that the ARMA-GARCH model for the PM_{10} variable has a higher RMSE value compared to the RF model. Furthermore, when examining the MAPE values, the RF technique outperforms the ARMA-GARCH model for both PM_{10} and $PM_{2.5}$ variables in the Seberang Jaya station, since it yields smaller MAPE values. However, in the case of the Seberang Perai station, the ARMA-GARCH model presents significantly smaller MAPE values for both PM_{10} and $PM_{2.5}$ variables in comparison to the RF method.

The MAPE values for the PM_{10} variable in Minden and Balik Pulau stations indicate that the ARMA-GARCH model yields larger values compared to the RF model. Additionally, for the $PM_{2.5}$ variable in Balik Pulau, the ARMA-GARCH model performs notably better, resulting in significantly lower MAPE values when contrasted with the RF model.

When focusing on the MAE values, it becomes apparent that the RF method excels, particularly for the PM_{10} and $PM_{2.5}$ variables, across all stations. The MAE values for the RF models are consistently significantly smaller than those of the ARMA-GARCH models, with the exception of the $PM_{2.5}$ variable for Seberang Perai, where the difference is slight. In summary, the RF method demonstrates superior forecasting performance for both PM_{10} and $PM_{2.5}$ variables across all stations. This conclusion is underpinned by the majority of RF models yielding smaller values across the three evaluation indices when compared to the ARMA-GARCH models.

9.5 CONCLUSION

This chapter focuses on one-day ahead forecasting of particulate matter concentrations at four monitoring stations in Penang, Malaysia, covering the period from 24 July 2017 to 31 December 2020. The dataset encompasses measurements of various air pollutants and meteorological variables. To analyse the dataset, two distinct techniques, RF and univariate ARIMA-GARCH, are employed and their accuracy in making predictions is compared. The feature importance assessment is carried out using the RF technique to identify the most influential explanatory variables in each model. Across all stations, a consistent pattern emerges, with PM_{10} and $PM_{2.5}$ consistently ranked as the two most crucial features for one-day ahead prediction of both PM_{10} and $PM_{2.5}$. However, it is important to note that the treatment of extreme values appears to have an impact solely on the Seberang Perai station. The results from the other stations consistently reinforce the same conclusion regarding the two most important features for one-day ahead prediction of both PM_{10} and $PM_{2.5}$.

In the comparison between the RF model and the univariate ARIMA-GARCH model on prediction, the best-performing models from each technique were assessed for all stations. The results indicate that, for the majority of cases, the RF models exhibit superior predictive abilities. This is evident as the RMSE, MAPE and MAE values are consistently significantly smaller in RF models compared to the univariate ARIMA-GARCH models. In some instances, no substantial differences in performance evaluation measures are observed. However, it is worth noting that for the Seberang Perai station, both the PM_{10} and $PM_{2.5}$ variables exhibited significantly smaller MAPE values in the univariate ARIMA-GARCH model compared to the RF model. Similarly, in the Balik Pulau station, the MAPE value of the univariate ARIMA-GARCH model is notably smaller than the MAPE values of the RF model for the $PM_{2.5}$ variable. In summary, within the scope of this study, the RF method demonstrates better performance in forecasting both PM_{10} and $PM_{2.5}$ variables across most stations. This conclusion is supported by the fact that the performance evaluation measures of RF models are consistently smaller than those of univariate ARIMA-GARCH models.

Having a good predictive model for particulate matter concentrations is of paramount importance, particularly for policymakers and public health officials. Accurate forecasts of air quality help policymakers make informed decisions about environmental regulations and interventions to mitigate pollution. These models provide critical insights into pollution sources and trends, enabling targeted efforts to improve air quality.

Moreover, the impact on human health cannot be overstated. Particulate matter, especially fine $PM_{2.5}$, is known to have detrimental effects on respiratory and cardiovascular health. A reliable predictive model empowers policymakers to take preventive actions and issue health advisories during periods of elevated pollution, potentially reducing the burden of respiratory illnesses and saving lives. It underscores the interconnectedness of environmental policy, public health and the well-being of communities.

9.6 ACKNOWLEDGEMENT

The authors would like to thank the Department of Environment Malaysia for providing the data for this study.

REFERENCES

Bollerslev, T. (1986). Generalized autoregressive conditional heteroskedasticity. *Journal of Econometrics*, *31*(3), 307–327.

Breiman, L. (2001). Random forests. *Machine Learning*, *45*(1), 5–32.

Engle, R. F. (1982). Autoregressive conditional heteroscedasticity with estimates of the variance of United Kingdom inflation. *Econometrica: Journal of the Econometric Society*, 987–1007.

Hussin, N. H., Yusof, F., Jamaludin, A. R., & Norrulashikin, S. M. (2021). Forecasting wind speed in Peninsular Malaysia: An application of ARIMA and ARIMA-GARCH models. *Pertanika Journal of Science & Technology*, *29*(1), 31–58.

Ivanov, A., Voynikova, D., Stoimenova, M., Gocheva-Ilieva, S., & Iliev, I. (2018, October 25). Random forests models of particulate matter PM10: A case study. In *AIP conference proceedings 2025*. AIP Publishing. https://doi.org/10.1063/1.5064879

Krueger, E. (2021). *Build better decision trees with pruning* [Online]. Retrieved August 17, 2022, from the World Wide Web: https://towardsdatascience.com/build-better-decision-trees-with-pruning-8f467e73b107

Kumar, U., & Jain, V. K. (2010). ARIMA forecasting of ambient air pollutants (O3, NO, NO2 and CO). *Stochastic Environmental Research and Risk Assessment*, *24*(5), 751–760.

Li, H. L., Zheng, R. J., Zheng, Q., Jiang, W., Zhang, X. L., Wang, W. M., Feng, X., Wang, K., & Lu, X. B. (2020). Predicting the number of visceral leishmaniasis cases in Kashgar, Xinjiang, China using the ARIMA-EGARCH model. *Asian Pacific Journal of Tropical Medicine*, *13*(2), 81–90.

Masseran, N. (2017). Modeling fluctuation of PM10 data with existence of volatility effect. *Environmental Engineering Science*, *34*(11), 816–827.

Radhika. (2020). *Out-of-Bag (OOB) score in the random forest algorithm* [Online]. Retrieved August 17, 2022 from the World Wide Web: www.analyticsvidhya.com/blog/2020/12/out-of-bag-oob-score-in-the-random-forest-algorithm/

Roy, S. S., Pratyush, C., & Barna, C. (2016, August 24–26). Predicting ozone layer concentration using multivariate adaptive regression splines, random forest and classification and regression tree. In *International workshop soft computing applications*, Arad, Romania.

Sharma, M., Jain, S., Mittal, S., & Sheikh, T. H. (2021). Forecasting and prediction of air pollutants concentrates using machine learning techniques: The case of India. In *IOP Conference Series: Materials Science and Engineering*, 24 October 2020. Rajpura, India.

Sutton, C. D. (2005). Classification and regression trees, bagging, and boosting. *Handbook of Statistics*, *24*, 303–329.

United States Environmental Protection Agency. (2022). *Particulate Matter (PM) basics* [Online]. Retrieved August 8, 2022, from the World Wide Web: www.epa.gov/pm-pollution/particulate-matter-pm-basics#PM

Wahir, N. A., Nor, M. E., Rusiman, M. S., & Gopal, K. (2018). Treatment of outliers via interpolation method with neural network forecast performances. In *Journal of physics: Conference series*, 28–29 October 2017. Hotel Katerina, Malaysia.

Wei, W. W. S. (2006). *Time series analysis: Univariate and multivariate methods* (2nd ed.). Pearson Addison Wesley.

Zaman, N. A. F. K., Kanniah, K. D., Kaskaoutis, D. G., & Latif, M. T. (2021). Evaluation of machine learning models for estimating PM2.5 Concentrations across Malaysia. *Applied Sciences*, *11*(16), 1–24.

CHAPTER 10

Applications of κ-Curves to Robot Trajectory Generation

Kenjiro T. Miura, Md Yushalify Misro, R. U. Gobithaasan, Dan Wang, Fengli Lan and Hirofumi Tamai

10.1 INTRODUCTION

κ-curve was proposed by Yan et al. (2017) and is characterised by several distinct features. First, the κ-curve is composed of multiple quadratic Bézier curve segments. Second, it is designed to pass through all input points, ensuring accuracy in following the specified path. Third, the curve aligns its curvature extreme points at both maxima and minima with the given input points, providing precise control over the shape. Last, it guarantees curvature continuity, also known as G^2 continuity, throughout the curve, except at inflection points where this continuity may be interrupted.

κ-curve is a planar curve and is intended for use in illustrations and other industrial designs. However, the third characteristic explains that if this curve is used for trajectory generation, dynamically critical points can be controlled. It has the potential to generate high-quality curves as trajectories for robots and so on. Therefore, in this chapter, we introduce the κ-curve with the intention of generating a trajectory, and extend it to a

 DOI: 10.1201/9781032678573-10

space curve, convert it into a G^2 curve at an inflection point, and use the εκ-curve, which makes it possible to control the extreme value of curvature. We will also introduce the spherical κ-curve, which can be used for end mill posture control of 5-axis machining tools and nozzle control of 3D printers.

10.2 κ-CURVE

Figure 10.1 shows an example of a κ-curve. The κ-curve is defined as segments of quadratic Bézier curves, and each segment can have at most one extreme value of signed curvature. The viewer has a tendency to pay attention to the point where the curvature reaches the extrema (Attneave, 1954). The κ-curve is effective for creating illustrations and is implemented as a curvature tool in Adobe Illustrator and Photoshop.

Figure 10.2 is an example that clearly shows the characteristics of the κ-curve, and a curve representing the curvature at that point is drawn outside the curvature depending on the sign of the signed curvature. We can see the absolute value of curvature is large where it spreads out greatly relative to the curve, and the extreme value of curvature is taken at the input point indicated by ○. Adjacent segments are connected with curvature continuity at points where the tangent vector to the outer curve is discontinuous. Also, the κ-curve is inflected at the point where the outer curve is discontinuous. Although the curvature is discontinuous at

FIGURE 10.1 Example of κ-curve (input point: ○).

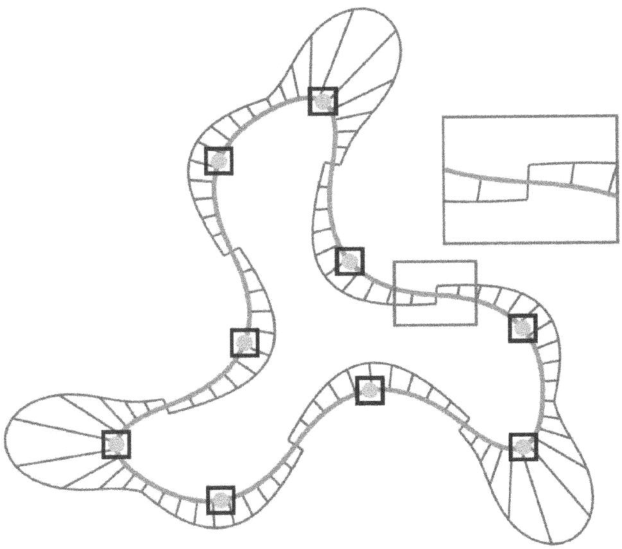

FIGURE 10.2 Characteristics of κ-curve.

the inflection point, the absolute values of the signed curvatures before and after are the same. In addition to closed curves, open curves can also be generated. The illustration shown in Figure 10.2 consists of one closed curve, but Figure 10.1 is a combination of multiple closed and open curves.

10.3 EXTENSIONS OF κ-CURVE

Although the κ-curve has good properties as a trajectory, it cannot be said to be desirable as a trajectory in the following two points.

1. It is only a plane curve and cannot represent spatial curves.
2. The curvature is discontinuous at the inflection point.

Three controls are used to define each segment before and after the connection point of two segments, as shown in Figure 10.3, so that the number of points that take the extreme value of curvature for each segment is at most one. To solve the first problem, we allow points to lie in different planes. That is, in Figure 10.3, control points $C_{i,0}$, $C_{i,1}$ and $C_{i,2}$ exist in a certain plane, and $C_{i+1,0} = C_{i,2}$, $C_{i+1,1}$ and $C_{i+1,2}$ exist in a different plane. Therefore, there is a "twisted" connection at $C_{i+1,0} = C_{i,2}$.

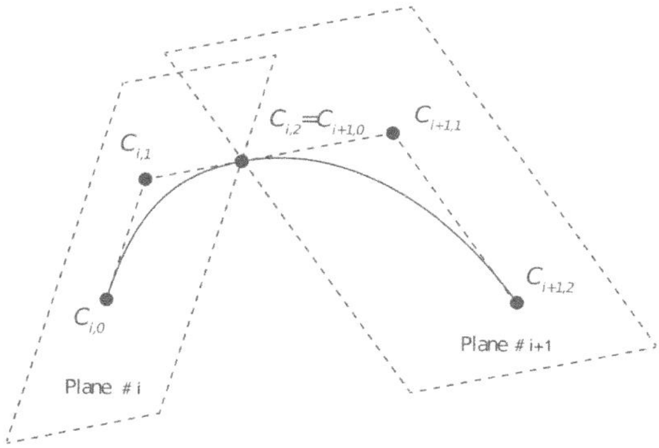

FIGURE 10.3 Space curve conversion of κ-curve.

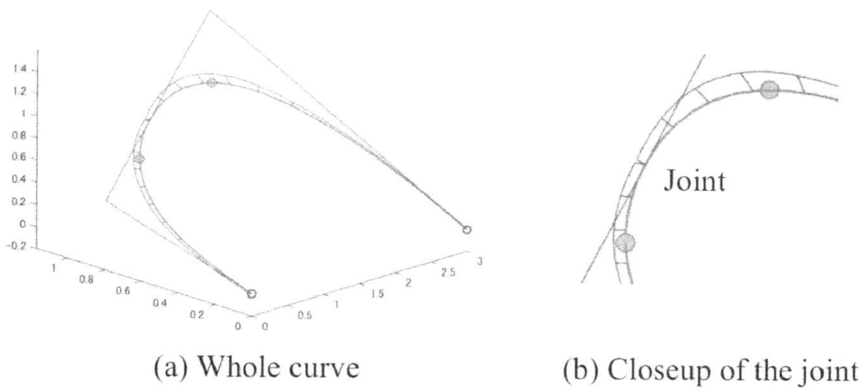

FIGURE 10.4 κ-space curve. (a) Whole curve. (b) Closeup of the joint.

Generation of such a curve can be achieved by a simple extension of the κ-curve generation algorithm, but as shown on the left in Figure 10.4, the input point is the extreme value of curvature, but as shown on the right in Figure 10.4, Although the magnitude of the curvature is the same at the connection point, the Frenet frames of the previous and next curve segments are different, and G^2 continuity is not guaranteed, but only G^1 continuity (tangential continuity) is guaranteed. Therefore, we solve this problem as well as the second problem.

To do this, replace these segments with a quartic Bézier curve that satisfies G^2 continuity at the end points as demonstrated in Figure 10.5. This replacement ensures G^2 continuity throughout the curve even if the connection point of the segment is an inflection point. Therefore, both problems are solved at the same time.

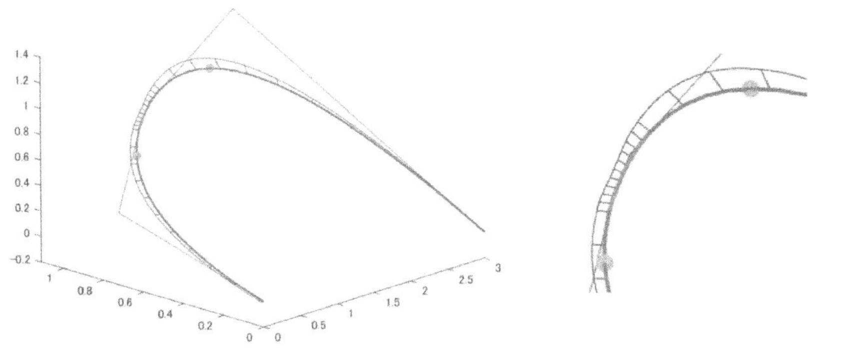

(a) Whole curve (b) Closeup of the joint

FIGURE 10.5 Partial replacement by a quartic Bezier curve. (a) Whole curve. (b) Closeup of the joint.

10.4 εκ-CURVE

εκ-Curve is an abbreviation for extended κ-curve, and by increasing the degree of the segments of the κ-curve, the degree of freedom of each curve segment is increased by 1, making it possible to control the value of the extreme value of curvature. Considering trajectory generation, being able to control the value of the extreme value of curvature also makes it possible to decrease curvature of the curve, which can improve the moving speed of objects and end effectors such as robot hands. Yan et al. (2019) increases the degree of freedom by one by rationalising the curve segment, but our proposed method can basically be applied to all curve types whose degree can be increased. The idea of the εκ-curve is that if the degree of freedom of the quadratic Bézier curve is increased by one to make it cubic, the degree of freedom of the x and y coordinates of the additional control points will be added, so the degree of freedom will increase by 2. There is one more extra degree of freedom than necessary. Therefore, the constraint condition shown in Figure 10.6 is applied to the second and third control points of the cubic Bézier curve.

In other words, let the three control points that define the quadratic Bézier curve be Q_0, Q_1 and Q_2, and let the control points of the cubic Bézier curve that has been raised to that degree be P_0, P_1, P_2 and P_3. Since the starting and ending points are to match, set $Q_0 = P_0$, $Q_2 = P_3$. Using the fixed constant a, the second control point P_1 is the point that

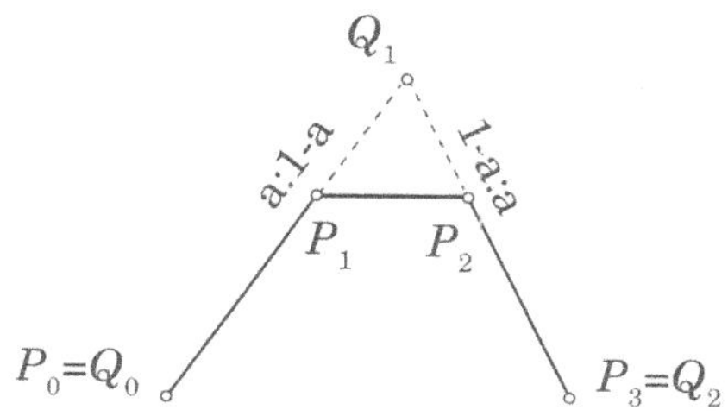

FIGURE 10.6 Constrained cubic Bézier curve.

internally divides side Q_0Q_1 into $a:1-a$, and the third control point P_2 is the point that internally divides side Q_1Q_2 into $1-a:a$. Expressed mathematically,

$$P_1 = (1-a)Q_0 + aQ_1 \tag{10.1}$$

$$P_2 = aQ_1 + (1-a)Q_2 \tag{10.2}$$

Note that line segment P_0P_3 and line segment P_1P_2 are parallel. The constrained cubic Bézier curve with $a = 2/3$ is identical to a quadratic Bézier curve. Another point to note is that even when using a cubic Bézier curve, each segment must have at most one curvature extremum, and care must be taken about the range of values of a. Miura et al. (2022c) prove that a constrained cubic Bézier curve has at most one curvature extremum when $2/3 \leq a \leq 1$. Figure 10.7 shows a comparison of the κ-curve and εκ-curve.

In Figure 10.7, the κ-curve is shown on the left, and the εκ-curve for various values of a is shown on the right. By increasing the value of a, the value of the curvature extrema increases, and the curve becomes sharper at the input point.

In summary, the εκ-curve exhibits several key characteristics. It inherits the foundational properties of the κ-curve, ensuring robust and reliable performance. The εκ-curve is versatile, being applicable to various types of curves, including polynomials, rational expressions and trigonometric curves. Notably, it does not require an increase in input points to control curvature extreme values, streamlining the design process. Additionally, the εκ-curve can be controlled both globally and locally, as demonstrated in Figure 10.8. Its processing speed is comparable to that of κ-curves, enabling interactive curve generation.

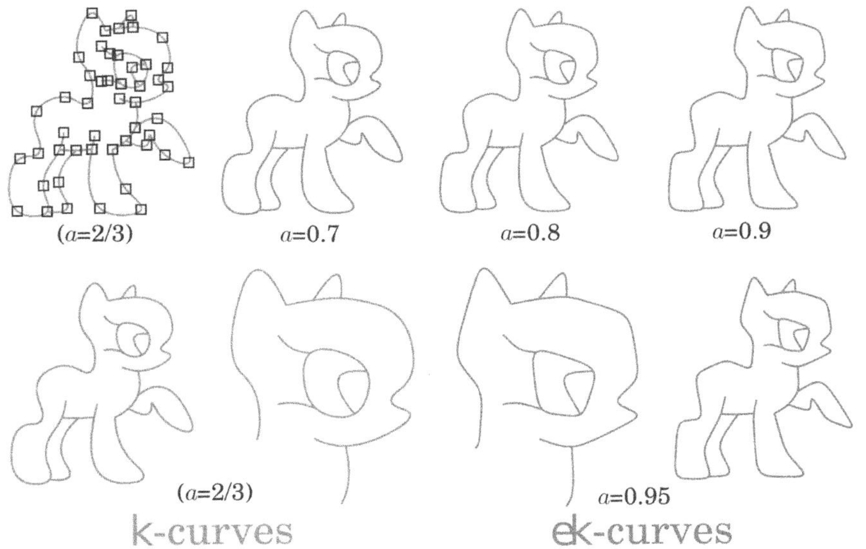

FIGURE 10.7 Comparison of κ-curve and εκ-curve.

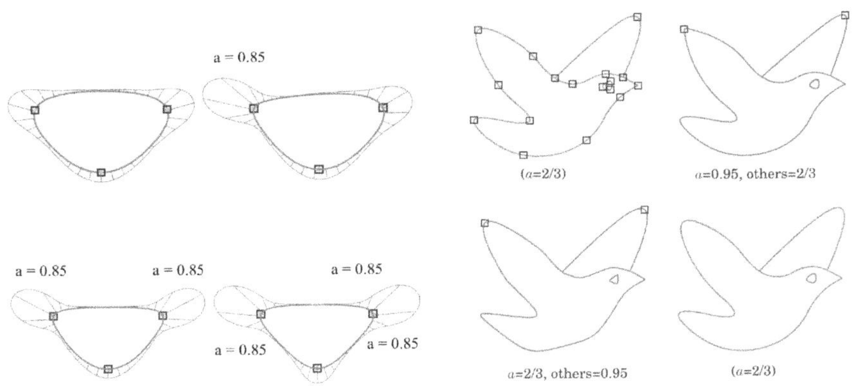

FIGURE 10.8 Local deformation examples. (a) First example. (b) Second example.

10.5 SPHERICAL κ-CURVE

With the development of micro-nanotechnology, manufacturing has become more complex and sophisticated, and there is a strong desire for higher precision and efficiency in machining tools and 3D printers.

However, at present, there is a lack of a sophisticated formulation for describing and controlling the position and posture of a controlled object, which hinders the development of advanced software. Therefore, as such

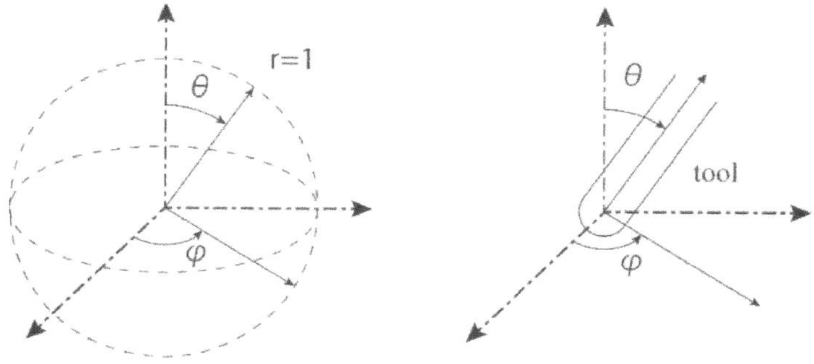

(a) Polar coordinate $r = 1$ (b) Specification of tool posture

FIGURE 10.9 Specifying posture using a curve on the unit sphere. (a) Polar coordinate $r=1$. (b) Specification of tool posture.

a formulation, we define a new κ-curve on the unit sphere—a spherical κ-curve—and propose that the spherical κ-curve be used for attitude control (Miura et al., 2022b).

The spherical κ-curve can specify passing points on the unit sphere and takes the extreme value of curvature at those passing points. By reducing its extreme value, it is possible to reduce the acceleration of posture control and the force required for it. There are two types of expressions: one based on rational quadratic equations and the other based on trigonometric functions, but both are low-order and can efficiently optimise posture control.

Controlling the position and posture of the end mill of a 5-axis machining tool and the nozzle of a 3D printer with high accuracy and high quality leads to higher precision and higher-speed machining, and higher precision and higher rigidity of objects produced by 3D printers.

When using the spherical κ-curve, considering that only the direction is specified, S^2 (unit sphere of cubic space: see Figure 10.9 is used as the generation space instead of SO^3 (third-order special orthogonal group; Miura, 2000). Therefore, for example, controlling the axial direction of a ball end mill in subtractive manufacturing or the direction of the injection nozzle of a 3D printer in additive manufacturing, we can use the spherical κ-curve.

10.5.1 Spherical Bézier Curve

We define a spherical Bézier curve by applying the de Casteljau algorithm using the great arc, which is a geodesic curve on the spherical surface, as a curve element. There are two possible ways to represent a great circle (Miura et al., 2022a):

1. Circular arc expressed using a rational quadratic Bézier curve.
2. An arc expressed using trigonometric functions.

In this chapter, spherical Bézier curves using rational quadratic Bézier curves are called B-type (Bézier-type) curves, and spherical Bézier curves using trigonometric functions are called T-type (trigonometric-type) curves.

10.5.2 B-Type and T-Type Spherical Bézier Curves

Consider generating a great circular arc connecting two points P_0 and P_1 given on the unit sphere. If the two points are not mutually poles, two great arcs can be generated, but a shorter great arc should be generated. In B-type curves, this great arc $B_a(t)$ is expressed by the following formula:

$$B_a(t; P_0, P_1) = \frac{(1-t)^2 P_0 + 2(1-t)twP_{01} + t^2 P_1}{(1-t)^2 + 2(1-t)tw + t^2}$$

Here, if the angle between P_0 and P_1 is $\theta = \cos^{-1}(P_0 \cdot P_1)$,

$$P_{01} = \frac{P_0 + P_1}{2\cos^2 \frac{\theta}{2}}, w = \cos\frac{\theta}{2}$$

On the other hand, T-type curves use trigonometric functions in the definition equation to define a great arc as

$$T_a(t; P_0, P_1) = \frac{P_0 \sin(\theta(1-t)) + P_1 \sin(\theta t)}{\sin\theta}$$

Note that spherical κ-curves can be defined for both B-type and T-type curves.

10.5.3 Generation of Spherical κ-Curves

Figure 10.10 shows an example of a spherical κ-curve generated by optimising for the input points to pass through geodesic curvature. The curve is optimised so that at the input points the geodetic curvatures (or

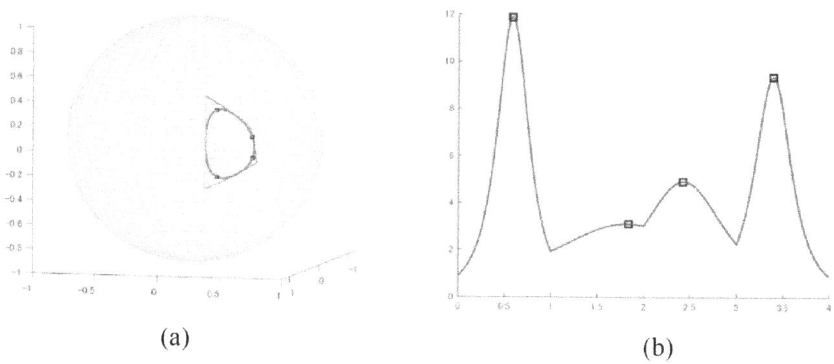

FIGURE 10.10 (a) Spherical κ-curve (b) Geodesic curvature distribution.

their absolute values) of two adjacent spherical quadratic Bézier curves match and the input point becomes the extreme value of the geodetic curvature.

In the example in Figure 10.10(a), a closed curve is generated on the unit sphere for four input points on that sphere. The right side of Figure 10.10(b) depicts the geodesic curvature of the generated curve, and the bending points in the graph indicate the connection positions of the quadratic spherical Bézier curve segments. The symbol □ indicates the position of the input point, and it can be seen from this graph that the geodesic curvature of each segment takes an extreme value at that point.

10.6 CONCLUSIONS

In this chapter, we have introduced the κ-curve with the intention of generating a trajectory, and also extended it to a space curve, converted it to G^2 at an inflection point, and also used the εκ(epsilon kappa)-curve, which allows us to control the extreme value of curvature. We also introduced the spherical κ-curve, which can be used for mill posture control of 5-axis machining tools and nozzle control of 3D printers. We omitted the explanation of the details of curve generation algorithm. We would like to ask readers interested in curve generation algorithms to refer to the references.

Research on κ-curves, especially spherical κ-curves, is still insufficient, and further research is expected in the future. Particularly, no literature on its application to trajectory generation is available, so we would like to continue such research in the future and contribute to manufacturing by implementing it on actual machines.

10.7 ACKNOWLEDGEMENT

Part of this chapter was supported by JSPS Grant-in-Aid for Scientific Research (B) 19H02048, JST CREST JPMJCR1911, and Kyushu University IMI short-term joint research project "New Developments in Discrete Differential Geometry: From Theory to Practice" in 2019.

REFERENCES

Attneave, F. (1954). Some informational aspects of visual perception. *Psychological Review, 61*(3), 183. https://psycnet.apa.org/doi/10.1037/h005466

Miura, K. T. (2000). Unit quaternion integral curve: A new type of fair free-form curves. *Computer Aided Geometric Design, 17*(1), 39–58. https://doi.org/10.1016/S0167-8396(99)00038-2

Miura, K. T., Gobithaasan, R. U., Inui, M., Sekine, T., & Usuki, S. (2022a). Degree elevation of spherical Bezier curve and its geodesic curvature at its end points. *Japan Society of Precision Engineering, Spring conference, in Japanese*. https://doi.org/10.11522/pscjspe.2022S.0_333

Miura, K. T., Gobithaasan, R. U., Inui, M., Sekine, T., & Usuki, S. (2022b). Spherical κ-curve. *Japan Society of Precision Engineering, Fall conference, in Japanese*. https://doi.org/10.11522/pscjspe.2022A.0_357

Miura, K. T., Gobithaasan, R. U., Salvi, P., Wang, D., Sekine, T., Usuki, S., Inoguchi, J. I., & Kajiwara, K. (2022c). εκ-Curves: Controlled local curvature extrema. *The Visual Computer, 38*(8), 2723–2738. https://doi.org/10.1007/s00371-021-02149-8

Yan, Z., Schiller, S., & Schaefer, S. (2019). Circle reproduction with interpolatory curves at local maximal curvature points. *Computer Aided Geometric Design, 72*, 98–110. https://doi.org/10.1016/j.cagd.2019.06.002

Yan, Z., Schiller, S., Wilensky, G., Carr, N., & Schaefer, S. (2017). κ-Curves: Interpolation at local maximum curvature. *ACM Transactions on Graphics (TOG), 36*(4), 1–7. https://doi.org/10.1145/3072959.3073692

CHAPTER 11

Solving the Global Path Planning of an Unmanned Ground Vehicle Using a Clustered Travelling Salesman Path Problem Approach

Veekeong Saw, Amirah Rahman and Wen Eng Ong

11.1 INTRODUCTION

An unmanned ground vehicle (UGV) is a robotic military vehicle with nobody on board. A UGV replaces the direct participation of combatants in military operations, which helps in reducing injuries and casualties. An important aspect considered in the development of UGV is its path navigation capability. As the UGV is driverless, the navigation process of

DOI: 10.1201/9781032678573-11

an UGV should require minimal human assistance and intervention. The UGV navigation process is divided into two types: local path planning (LPP) and global path planning (GPP). The LPP process maintains vehicle stability and safety in real time using local navigator sensors, whereas the GPP process searches for the most suitable travel route using terrain data before the UGV departs (Giesbrecht, 2004).

In general, the objective of the UGV GPP process is to complete a given task efficiently while minimising resource consumption. Note that the term "resource" may refer to non-physical attributes such as threat exposure, signal jamming attack and terrain roughness. From a broader perspective, a GPP process can be modelled as a shortest path problem with constraints. The problem varies based on the GPP mission assigned, which eventually requires different types of algorithm implementations. For instance, Han et al. (2014) implemented a label-correcting algorithm for solving a multi-criteria GPP problem, and Bae et al. (2015) presented a two-phase algorithm for solving a UGV reconnaissance problem. More recent work related to UGVs and path planning include Wei et al. (2018), Thoresen et al. (2021) and Baras and Dasygenis (2023).

In this chapter, we consider the UGV multi-criteria GPP problem with multiple visiting checkpoints, which is an extension of the study by Han et al. (2014). Han et al. (2014) considered the GPP problem for a pair of start-destination points on terrain with three attributes: traverse time, risk level and signal jamming level. They proposed a label correcting algorithm to find the Pareto optimal solution path. They also introduced a weighting approach that returns a solution path in a large-sized terrain map with a shorter time. However, in a real-life situation, a GPP operation might involve multiple visiting checkpoints instead of two checkpoints only. A quick solution method to find the travel path in such situations is needed.

The GPP problem in this chapter is a multicriteria optimisation problem (MOP), which deals with optimisation problems with multiple conflicting benefits. In MOP, there usually does not exist an ideal solution which optimises all attributes simultaneously. This situation calls for the use of Pareto optimality, where a set of maximal promising solutions is found such that further improvement towards an attribute will result the degradation of at least one of the other attributes (Luc, 2008). A simple approach to solving MOP is by converting it into a single-criteria optimisation problem (SOP) using the weighted sum method, ε-constraint method and ideal point method (Ehrgott, 2008; Marler & Arora, 2010).

From the weighted sum method, Przybylski et al. (2010) and Benson and Sun (2002) introduced the weight set decomposition algorithm to generate all Pareto-optimal solutions by assigning the weights systematically.

Another aspect of the extended GPP problem is that the UGV departs and ends at distinct checkpoints while visiting all other checkpoints at least once. The goal is to minimise the total cost over all attributes. The same terrain attributes as in Han et al. (2014) are assumed. Note that these attributes often conflict with each other. For example, an open space takes less time to traverse, but it increases the risk of exposure. The configuration space for this problem is constructed using the cell decomposition method. This is because the grid is the most suitable representation method for storing the configuration data of the entire terrain used during the computation process.

The extended GPP problem resembles the travelling salesman path problem (TSPP), which is a variant of the TSP. The only difference between the TSP and the TSPP is that the objective of TSPP is to find a Hamilton path instead of Hamilton cycle. The TSPP has been implemented for GPP purposes. For example, Saw et al. (2017a) discussed the procedure to solve the grid-based single-criteria TSPP, and Saw et al. (2017b) proposed the two-phase algorithm to solve the multicriteria GPP problem by solving the TSPP inside the complete graph modelled from the GPP problem.

In TSP-based problems, the computational effort increases rapidly with the problem size and complexity. This even occurs on approximation algorithms which find suboptimal solutions quickly. On a terrain, vertices that stay close to each other can be grouped into a cluster. Consequently, a large-sized TSP-based problem can be transformed into a clustered TSP. Clustered TSP is another well-known variant of TSP with the same objective as TSP, except that the path visits checkpoint within a cluster before moving on to the next cluster. Several discussions regarding clustered TSP can be found in Sheng et al. (2005), Mestria (2018) and Lu et al. (2022). Here, we adapt the work from Sheng et al. (2005) and Saw et al. (2017b) to develop a solution scheme for solving our GPP problem. We consider the GPP problem as a two-level clustered TSPP.

The chapter is organised as follows. Section 11.2 discusses basic notions, terminologies and the description of the GPP problem along with its mathematical formulation. In section 11.3, the solution scheme for the solution path for the GPP problem and the computational implementations are described, followed by the results and discussions in section 11.4. Section 11.5 concludes this chapter with suggestions for possible future research.

11.2 BASIC NOTIONS, TERMINOLOGIES AND PROBLEM DESCRIPTION

In this section, the formulation of the GPP problem is discussed. Following that, some basic definitions and terminologies that are used throughout this chapter are outlined.

11.2.1 Modelling the Problem

In this study, we consider the GPP problem for a UGV which consists of multiple visiting checkpoints. The UGV departs and ends at uniquely specified checkpoints while visiting other checkpoints known as intermediate checkpoints at least once, disregarding the visiting order. The UGV should complete the visit with minimal resource consumption. In this problem, several assumptions are considered:

1. The UGV is in good enough condition such that it can travel the entire journey without stopping for servicing or refuelling.
2. The UGV can perform the LPP process (obstacle avoidance, stability maintenance and changing direction) autonomously.
3. The cost assigned on the terrain does not reflect the actual attribute level. It serves as the estimated amount which will be incurred when the UGV traverses into that region.
4. The UGV may traverse through the same location more than once if necessary. The attribute cost will be added to the total path cost every time the UGV revisits that region.

In the mathematical context, the objective of the GPP problem is to search for a Pareto-optimal *st*TSPP solution path with uniquely given endpoints on a terrain that has been inscribed into a grid. An example of an $M \times N$ grid is shown in Figure 11.1.

In the $M \times N$ grid, the coordinate system is configured with the origin located at the top left cell; the horizontal axis goes from 0 to $N-1$, and the vertical axis goes from 0 to $M-1$. The coordinate (x, y) for a cell is assigned with row number x first, followed by the column number y. A UGV in the cell can travel to any of the eight adjacent cells, except for the boundaries of the grid as shown in Figure 11.1. This simulates the freedom of the UGV when moving in a real-world environment.

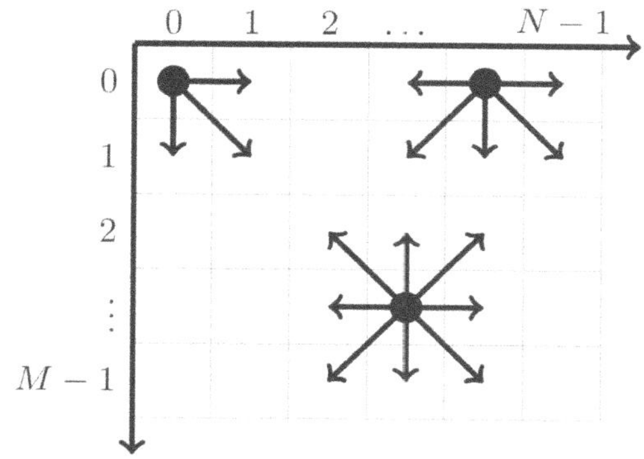

FIGURE 11.1 A grid with M rows and N columns. The bold arrows show the accessible movement of UGV while situated at different cells.

The grid is modelled into a weighted grid graph $G=(V,E,c)$ where vertex set V represents the cells and $E=\{e_{ij}|v_i,v_j\in V\}$ represents the set of undirected edges connecting two adjacent cells. The cost function $c:V\rightarrow\mathbb{R}^3$ maps every vertex $v_i\in V$ to a 3-dimensional cost vector $c(v_i)=(c_i^1,c_i^2,c_i^3)$, where c_i^1, c_i^2 and c_i^3 resemble the traverse time, risk level and jamming level at vertex v_i, respectively. The user then provides a set of $n+2$ visiting checkpoints $\Lambda=\{v_s,v_1,v_2,\cdots,v_n,v_t\}\subsetneq V$. The vertex v_s and v_t are the origin and destination checkpoint, respectively, while $v_1,v_2,\cdots,v_n$ are intermediate checkpoints.

Let x_i be defined by

$$x_i=\begin{cases}0, \text{ if vertex } v_i \text{ is traversed in the path,}\\ 1, \text{ otherwise.}\end{cases}$$

Let $Suc(v_i)$ and $Pre(v_i)$ be defined by successor set and predecessor set of vertex v_i such that $Suc(v_i)=\{v_j|e_{ij}\in E\}$ and $Pre(v_i)=\{v_j|e_{ji}\in E\}$ respectively. Also, let S be a nonempty subset of V. The mathematical formulation for the GPP problem is as follows:

$$\text{Minimise} \quad \sum_{v_i\in V} x_i c_i^k, \qquad k=1,2,3, \tag{11.1}$$

$$\text{subject to} \quad \sum_{v_j\in Suc(v_i)} x_j \geq 1, \qquad \forall v_i\in\Lambda\setminus\{v_t\}, \tag{11.2}$$

$$\sum_{v_j \in Suc(v_t)} x_j \geq 0, \tag{11.3}$$

$$\sum_{v_j \in Pre(v_i)} x_j \geq 1, \forall v_i \in \Lambda \setminus \{v_s\}, \tag{11.4}$$

$$\sum_{v_j \in Pre(v_s)} x_j \geq 0, \tag{11.5}$$

$$\sum_{v_j \in Suc(v_i)} x_j + \sum_{v_j \in Pre(v_i)} x_j = 2k_i, \quad k_i \in \mathbb{Z}^{0+}, \forall v_i \in \Lambda \setminus \{v_s, v_t\}, \tag{11.6}$$

$$\sum_{v_j \in Suc(v_i)} x_j + \sum_{v_j \in Pre(v_i)} x_j = 2k_i + 1, k_i \in \mathbb{Z}^{0+}, \forall v_i \in \{v_s, v_t\}, \tag{11.7}$$

$$\sum_{v_i \in S} \sum_{\substack{v_j \in \\ Suc(v_i) \cap S'}} x_j \geq 1, \ \forall S \subsetneq V, S \neq \varnothing_i, \tag{11.8}$$

$$x_j \in \{0,1\}, \forall v_i \in V. \tag{11.9}$$

Objective function (1) minimises the three attributes (traverse time, risk level and jamming level), represented by c_i^1, c_i^2 and c_i^3 for each vertex i. Constraint (2) ensures the path leaves every checkpoint in $\Lambda \setminus \{v_t\}$ at least once. Constraint (3) states the path may end at v_t or pass through v_t multiple times if necessary. Constraint (4) ensures the path enters every checkpoint in $\Lambda \setminus \{v_s\}$ at least once, Similarly, constraint (5) states that the path may begin at v_s or pass through v_s multiple times if necessary. Constraints (6) and (7) preserve the flow conservation throughout the path. Note that since every vertex can be traversed more than once, then for any vertex v_i in (6) (except v_s and v_t), the RHS is written as an even number $2k_i$ instead of 2. Similarly, endpoints v_s and v_t can also be traversed more than once. However, the journey must also start and end at v_s and v_t respectively, which contributes an additional degree for the vertices. Thus, in constraint (7) the RHS is written as an odd number $2k_i + 1$ instead of 1. Constraint (8) prevents the solution path from forming subtours. Finally, constraint (9) insists that the decision variables are binary.

11.2.2 Graph Theory

In the grid graph G, a path that connects vertex v_p to v_q, denoted by π_{pq}, is a sequence of vertices that the UGV visits, so we write $\pi_{pq} = \langle v_p, \cdots, v_q \rangle$. A path that begins and ends at the same vertex v_p is known as a cycle, denoted by $Cyc(v_p)$. Given two paths π_{pq} and π_{qr} where $p \neq q \neq r$, a concatenation operator $\oplus$ joins the paths π_{pq} and π_{qr} to form the path $\pi_{pr} = \langle v_p, \cdots, v_q, \cdots, v_r \rangle$ such that $\pi_{pr} = \pi_{pq} \oplus \pi_{qr}$. A path π_{ij} that connects two checkpoints $v_i, v_j \in \Lambda$ is known as partial path and is denoted by $\overline{\pi}_{ij}$. On the other hand, a path that connects v_s and v_t while passing through all intermediate points is known as a complete path and is denoted by π^*_{st}. Note that a complete path π^*_{st} with $n+2$ checkpoints results from the concatenation of $n+1$ partial paths such that $\pi^*_{st} = \overline{\pi}_{si_1} \oplus \overline{\pi}_{i_1 i_2} \oplus \cdots \oplus \overline{\pi}_{i_{n-1} i_n} \oplus \overline{\pi}_{i_n t}$. A complete graph K_n is a simple undirected graph with n vertices that have $\binom{n}{2} = \frac{n(n-1)}{2}$ edges connecting all pair of vertices. A path that passes through all vertices in K_n is known as a Hamilton path. Similarly, a cycle that passes through all vertices in K_n is known as a Hamilton cycle.

To calculate the cost of a path π_{pq}, we define the path cost $c(\pi_{pq})$ as the cumulative costs of all the traversed cells such that $c(\pi_{pq}) = \sum_{v_i \in \pi_{pq}} c(v_i)$. If a cell was visited more than once, then the cost will be added accordingly. Similarly, the cost of a cycle, partial path and complete path, denoted by $c(Cyc(v_p))$, $c(\overline{\pi}_{ij})$ and $c(\pi^*_{st})$, respectively, are calculated in the same way.

11.2.3 Multicriteria Optimisation Problem to Single-Criteria Optimisation Problem

As our GPP problem resembles a 3-dimensional MOP, every complete solution path π^*_{st} corresponds to a unique solution cost vector $c(\pi^*_{st})$, which can be mapped as a point on the $\mathbb{R}^3$ solution space. The x, y and z-axes of the solution space represent the cost for attributes 1, 2 and 3, respectively.

In practical MOP, the number of Pareto-optimal solutions is usually very large. One of the simplest ways to solve it is through the weighted sum method. A weight $\mathbf{w} = (w_1, w_2, w_3)$ that reflects the level of importance of each attribute is introduced to aggregate the MOP into an SOP. The weighted cost $w_k c^k$ is later summed up and minimised. This converts the MOP into the following SOP:

$$\text{Minimise}\left\{\sum_{k=1}^{3} w_k c^k \,\middle|\, \sum_{k=1}^{3} w_k = 1,\ w_k > 0 \forall k\right\} \tag{11.10}$$

After transforming the MOP into an SOP, simple algorithms can be implemented onto the SOP directly. The optimal solution obtained from the SOP corresponds to a Pareto-optimal solution in the MOP. To ensure Pareto-optimal solutions always exist, the condition $w_k > 0$ in (10) prevents the objective hyperplane in the solution space from being parallel to the solution axis in the solution space which results in dominated solutions.

In the solution space, we define the ideal point as $c^* = \left(c_1^*, c_2^*, c_3^*\right)$, where c_k^* is the lexicographically optimal complete path cost for attribute k alone. The ideal point resembles the most perfect situation that the GPP problem wishes to achieve. To measure the closeness of a complete solution path π_{st}^* from the ideal state, we measure the L_2 distance of the solution point $c\left(\pi_{st}^*\right) = \left(c^1, c^2, c^3\right)$ from the ideal point c^* in the solution space

$$\left[\sum_{k=1}^{3} \left|c^k - c_k^*\right|^2\right]^{\frac{1}{2}}, \tag{11.11}$$

where the solution path that has shortest L_2 distance from c^* is closest to the ideal state compared to other solution paths.

11.2.4 Point Clustering

Consider the checkpoint set Λ. We can partition Λ into $p > 3$ many non-intersecting subsets $\Lambda_1 = \{v_s\}, \Lambda_2, \cdots, \Lambda_p = \{v_t\}$ such that $\bigcup_{i=1,2,\cdots,p} \Lambda_i = \Lambda$. These partitions are known as clustered checkpoints or clusters in short. For any two clusters Λ_i and Λ_j, the distance between Λ_i and Λ_j, denoted by $d\left(\Lambda_i, \Lambda_j\right)$, is given by the minimal Euclidean distance between $v_i \in \Lambda_i$ and $v_j \in \Lambda_j$ in the grid such that

$$d\left(\Lambda_i, \Lambda_j\right) = \min_{v_i \in \Lambda_i, v_j \in \Lambda_j} \left\{\sqrt{\left(x_i - x_j\right)^2 + \left(y_i - y_j\right)^2}\right\}. \tag{11.12}$$

The radius of cluster Λ_k, denoted by $r\left(\Lambda_k\right)$, is given by the maximal Euclidean distance between the pairs of checkpoints in Λ_k in the grid such that

$$r\left(\Lambda_k\right) = \max_{v_i, v_j \in \Lambda_k} \left\{\sqrt{\left(x_i - x_j\right)^2 + \left(y_i - y_j\right)^2}\right\}. \tag{11.13}$$

Using these clusters, we construct a new graph $K_{cls} = \left(V_{cls}, E_{cls}, c_{cls}\right)$ such that

- Vertex set $V = \Lambda_1, \Lambda_2, \cdots, \Lambda_p$.
- Edge set $E_{cls} = \left\{ e_{ij}^{cls} \middle| e_{ij}^{cls} = arg\ min\left\{ d\left(\Lambda_i, \Lambda_j\right) \right\}, \Lambda_i, \Lambda_j \in V_{cls} \right\}$.
- Cost function $c_{cls} : E_{cls} \to \mathbb{R}$ maps every $e_{ij}^{cls} \in E_{cls}$ to their distance such that $c_{cls}\left(e_{ij}^{cls}\right) = d\left(\Lambda_i, \Lambda_j\right)$.

In K_{cls}, the intercluster path from cluster Λ_p to Λ_q denoted by π_{pq}^{cls} is a Hamilton path with the endpoints Λ_p and Λ_q. The cost of the intercluster path π_{pq}^{cls} is $c\left(\pi_{pq}^{cls}\right) = \sum_{e_{ij}^{cls} \in \pi_{pq}^{cls}} d\left(\Lambda_i, \Lambda_j\right)$. Note that every complete path π_{st}^{*} for the GPP problem corresponds to an intercluster path $\pi_{1,p}^{cls}$ with Λ_1 and Λ_p as its endpoints.

11.3 SOLUTION APPROACH

We consider the GPP problem as a two-level clustered TSPP which is solved in both global and local perspectives. In the global perspective, the shortest intercluster path $\pi_{1,p}^{cls}$ that visits every cluster is determined by solving the TSPP in graph K_{cls}. On the other hand, in the local perspective, the intracluster Hamilton path for each cluster is found by solving the TSPP for the checkpoints within each cluster. Finally, all intracluster Hamilton paths obtained are joined together according to sequence from the intercluster path to form a complete path. An example of a GPP problem considered as a two-level clustered TSPP is illustrated in Figure 11.2.

The solution scheme procedure is illustrated in Figure 11.3. The solution scheme is divided into two phases: preprocessing and solution path searching. In the preprocessing phase, the user provides the necessary inputs to prepare the grid graph and to determine the shortest intercluster checkpoint visiting sequence. In the solution path searching phase, the shortest intracluster TSPP solution path for each cluster is determined. These paths are combined to form the complete path. Finally, 2-opt improvement is conducted to improve the quality of the solution path obtained.

11.3.1 Preprocessing Phase

In the preprocessing phase, the grid map for different attributes, coordinates of visiting checkpoints and the weight assigned to each attribute

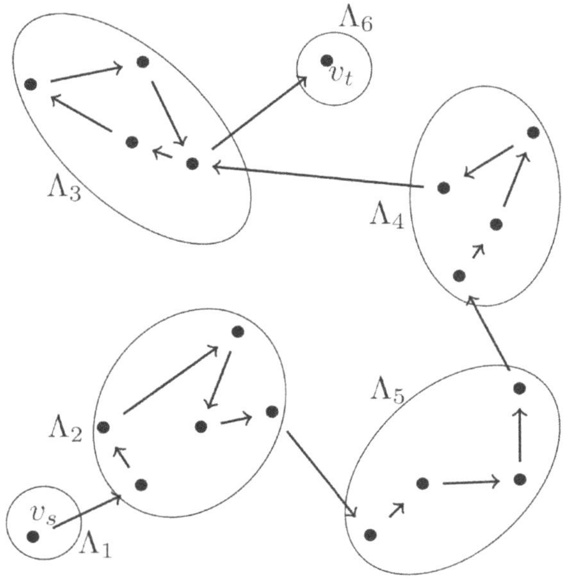

FIGURE 11.2 An example of TSPP considered as a two-level clustered TSPP.

are provided. These inputs are used to convert the multicriteria-based GPP problem into an SOP. By referring to the procedure flowchart given in Figure 11.3, Part (A) handles the preprocessing of the grid map, while Part (B) manages the clustering of checkpoints and intercluster path searching.

In Part (A), the configuration space is constructed from the terrain map using the cell decomposition method. The grid is modelled into a vertex-weighted grid graph $G = (V, E, c)$ with a 3-dimensional cost vector $c(v_i) = (c_i^1, c_i^2, c_i^3)$. Since the range of terrain data from multiple sources might vary, these attribute costs are normalised into a common range $[a, b]$ to produce normalised cost $\hat{c}(v_i) = (\hat{c}_i^1, \hat{c}_i^2, \hat{c}_i^3)$ such that

$$\hat{c}_i^k = a + \frac{c_i^k - c_{min}^k}{c_{max}^k - c_{min}^k} \times (b - a), \tag{11.14}$$

where c_{min}^k and c_{max}^k are the smallest and largest cell cost for attribute k, respectively. After normalising the cost on the graph, the cost $\hat{c}$ is aggregated with the weight $\mathbf{w}$ to form the combined weighted cost $c^{\mathrm{w}} = \Sigma_{K=1}^{3} w_k \hat{c}_k$. This reduces the grid graph G with 3-dimensional cost functions into $G' = (V, E, c^{\mathrm{w}})$ with a single dimensional cost function.

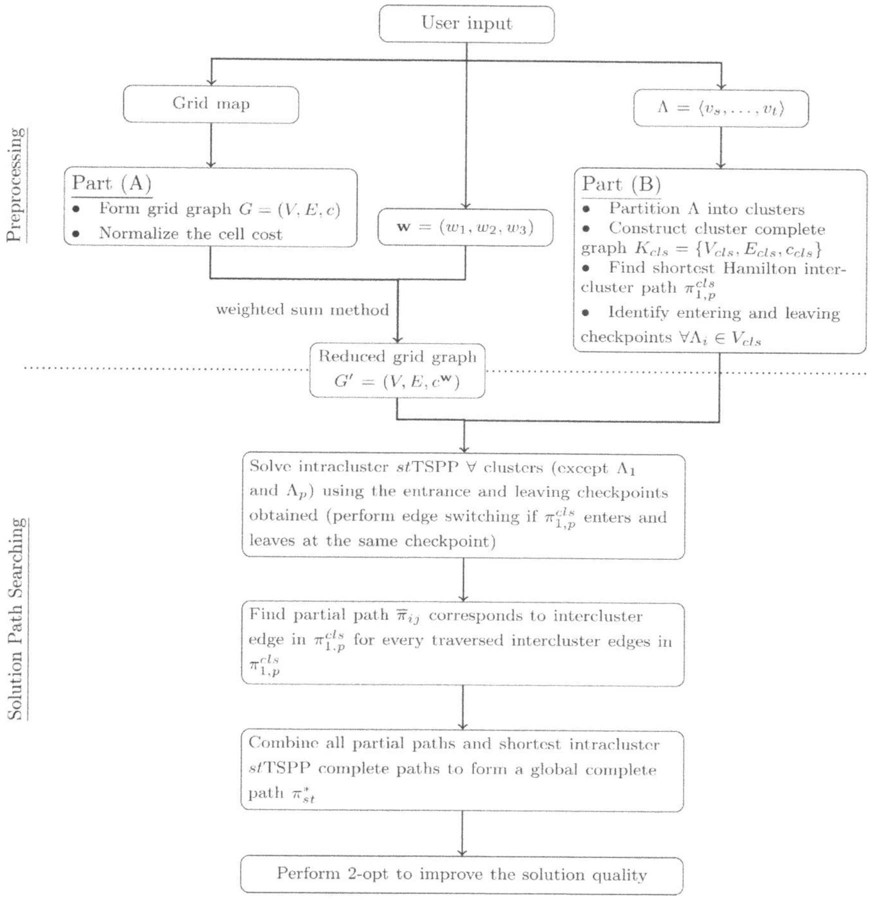

FIGURE 11.3 Procedure for the proposed solution scheme.

Part (B) begins with the partitioning of checkpoints Λ into clusters using *k*-Means clustering algorithm. Note that v_s and v_t are required to be in their own clusters which contain no other checkpoints. Thus, they are not considered while performing *k*-Means clustering. All remaining checkpoints are divided into $k-2$ clusters in total. The centroid for each cluster is calculated by averaging the x and y-coordinates of all the checkpoints. The algorithm for *k*-Means clustering for our GPP problem is presented in Algorithm 11.1.

Algorithm 11.1 *k*-Means Clustering Algorithm for the GPP Problem

Input: number of clusters *k*, checkpoint set Λ

Output: *k* many clusters $\Lambda_1, \Lambda_2, \cdots, \Lambda_k$

1: $\Lambda \to \Lambda \setminus \{v_s, v_t\}$

2: Assign $\Lambda_1 = \{v_s\}$ and $\Lambda_k = \{v_t\}$

3: **for all** remaining checkpoints in Λ **do**

4: Select $k-2$ many checkpoints as initial centroids

5: **repeat**

6: Form $\Lambda_2, \Lambda_3, \cdots, \Lambda_{k-1}$ by assigning each checkpoint to its nearest centroid

7: Recompute the centroid for each cluster

8: **until** all centroids are fixed

9: end for

After performing the *k*-Means clustering algorithm, we have the clusters $\Lambda_1, \Lambda_2, \cdots, \Lambda_p$ with v_s in the first cluster Λ_1 and v_t in the last cluster Λ_p. Using these clusters, the graph $K_{cls} = (V_{cls}, E_{cls}, c_{cls})$ is formed. From K_{cls}, the shortest intercluster path with endpoints Λ_1 and Λ_p is determined. This path represents the sequence for visiting the clustered checkpoints in the GPP problem. The entry and leaving checkpoints for each cluster are used as the origin and destination vertices when solving for the intracluster TSPP in each cluster.

11.3.2 Solution Path Searching Phase

Once the intercluster path $\pi_{1,p}^{cls}$ is determined, the intracluster TSPP for each cluster Λ_i can now be determined using the two-phase algorithm introduced by Saw et al. (2017b). The first step is to form a complete graph that connects all checkpoints within a cluster. This is done by searching for all partial paths that connect every pair of checkpoints in the cluster using a modified A* algorithm. Once the complete graph containing the clustered checkpoints with cost function c^w is formed, the shortest intracluster Hamilton path for each cluster can be found using repetitive nearest neighbour algorithm (RNNA; see Saw et al., 2017b).

Note that for an intercluster path $\pi_{1,p}^{cls}$ in K_{cls}, there are two possible situations that might occur:

1. The intercluster path $\pi_{1,p}^{cls}$ enters and leaves the intermediate clusters at different checkpoints (as in clusters Λ_2, Λ_4, and Λ_5 of Figure 11.2).

2. The intercluster path $\pi_{1,p}^{cls}$ enters and leaves the intermediate clusters at the same checkpoints (as in cluster Λ_3 of Figure 11.2).

For the first situation, the intracluster Hamilton path can be found directly because it resembles the TSPP with distinct entry and leaving checkpoints. In this situation, RNNA can be applied directly onto the TSPP. However, in the second situation, the cluster itself does not resemble a TSPP because $\pi_{1,p}^{cls}$ enters and leaves at the same checkpoint. Due to the triangle inequality, the shortest Hamilton cycle obtained will have a higher travel cost than the shortest intercluster Hamilton path obtained within the cluster. Thus, an edge-switching process needs to be done to ensure the path has different endpoints.

Suppose that $\pi_{1,p}^{cls} = \langle \Lambda_1, \cdots, \Lambda_{k-1}, \Lambda_k, \Lambda_{k+1}, \cdots, \Lambda_p \rangle$ is the shortest intercluster path with cluster Λ_k having the same entering and leaving checkpoint so that $|\Lambda_k| = s+1$. Let $\langle v_i, v_{n_1}, v_{n_2}, \cdots, v_{n_{s-1}}, v_{n_s}, v_i \rangle$ be the shortest Hamilton cycle obtained from cluster Λ_k. Also, suppose that v_x is the leaving checkpoint for cluster Λ_{k-1} and v_y is the entering checkpoint for cluster Λ_{k+1}. Thus, we have the intracluster Hamilton cycle $\langle \cdots, v_x, v_i, v_{n_1}, v_{n_2}, \cdots, v_{n_{s-1}}, v_{n_s}, v_i, v_y, \cdots \rangle$ for cluster Λ_k. The edge-switching process is conducted by examining the four possible alteration cases, and the one which gives the lowest total cost will be considered. These four possible path alterations are as follows:

1. Replace $\overline{\pi}_{n_s,i}$ and $\overline{\pi}_{i,y}$ into $\overline{\pi}_{n_s,y}$ to obtain the path segment $\cdots, v_x, v_i, v_{n_1}, v_{n_2}, \cdots, v_{n_{s-1}}, v_{n_s}, v_i, v_y, \cdots$ as illustrated in Figure 11.4.
2. Replace $\overline{\pi}_{x,i}$ and $\overline{\pi}_{i,n_1}$ into $\overline{\pi}_{x,n_1}$ to obtain the path segment $\cdots, v_x, v_{n_1}, v_{n_2}, \cdots, v_{n_{s-1}}, v_{n_s}, v_i, v_y, \cdots$ as illustrated in Figure 11.5.
3. Reverse the Hamilton cycle, replace $\overline{\pi}_{n_1,i}$ and $\overline{\pi}_{i,y}$ into $\overline{\pi}_{n_1,y}$ to obtain the path segment $\cdots, v_x, v_i, v_{n_s}, v_{n_{s-1}}, \cdots, v_{n_2}, v_{n_1}, v_y, \cdots$ as illustrated in Figure 11.6.
4. Reverse the Hamilton cycle, replace $\overline{\pi}_{x,i}$ and $\overline{\pi}_{i,n_s}$ into $\overline{\pi}_{x,n_s}$ to obtain the path segment $\cdots, v_x, v_{n_s}, v_{n_{s-1}}, \cdots, v_{n_2}, v_{n_1}, v_i, v_y, \cdots$ as illustrated in Figure 11.7.

After performing the edge-switching process, every intermediate cluster in the local intracluster TSPP will have the shortest Hamilton paths instead of Hamilton cycles. Once the shortest intercluster Hamilton

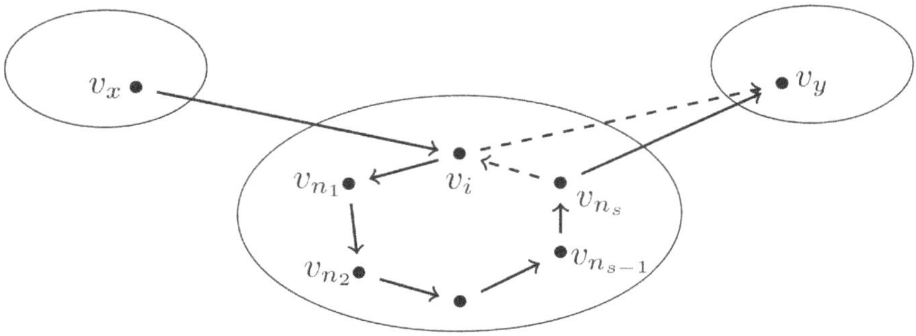

FIGURE 11.4 Path alteration case 1.

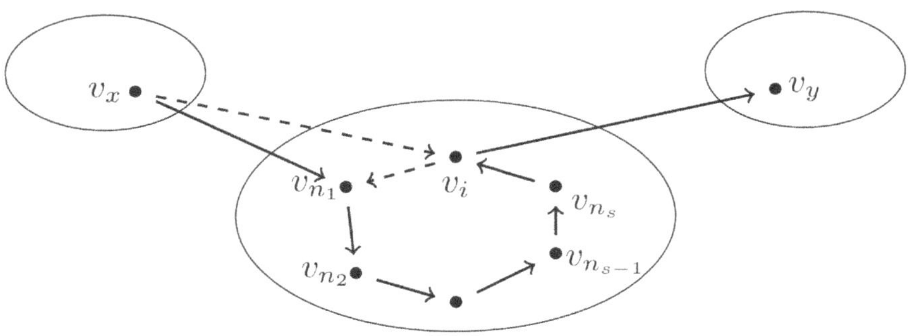

FIGURE 11.5 Path alteration case 2.

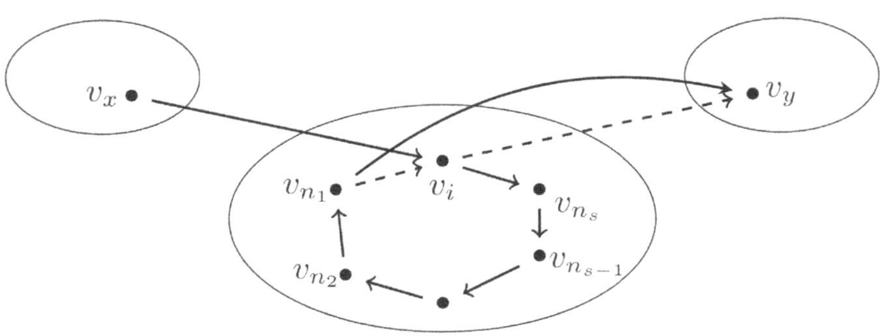

FIGURE 11.6 Path alteration case 3.

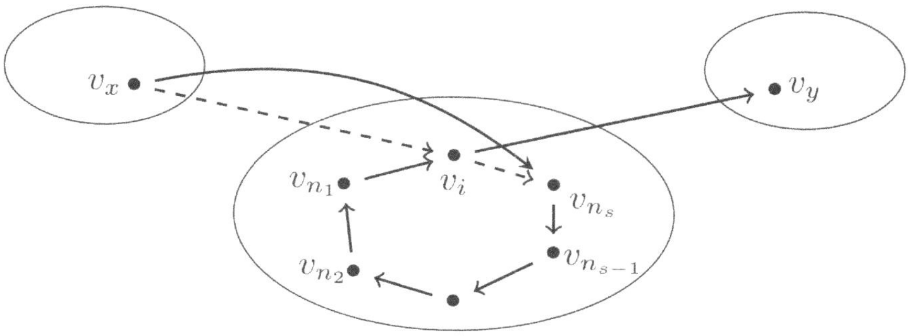

FIGURE 11.7 Path alteration case 4.

path for each cluster has been determined, the procedure continues with searching for the shortest partial paths that connect the clusters according to the sequence in $\pi_{1,p}^{cls}$ using a modified A* algorithm. These partial paths begin from the leaving checkpoint of the previous cluster and end at the entry checkpoint of the subsequent cluster. These partial paths are finally concatenated with the shortest intracluster Hamilton path obtained from each cluster appropriately to form a lowest cost global complete path π_{st}^*. In order to enhance the quality of solution path obtained from the previous steps, the global 2-opt process are performed. In this process, all pairs of interchanged edge costs in the π_{st}^* are examined. If a path with a lesser cost is found, the edge will be swapped. The process of 2-opt algorithm is shown in Algorithm 11.2.

Algorithm 11.2 2-Opt Algorithm

Input: initial complete solution path $\pi_{st}^* = \langle v_s, v_{i_1}, v_{i_2}, \cdots, v_{i_n}, v_t \rangle$

Output: new Hamilton path π_{st}^* with lower travel cost

1: **for all** $i = 1, 2, \cdots, n+1$ **do**

2: **for all** $j = 1, 2, \cdots, n+1$ **do**

// $e_{[k,k+1]}$: the edge connecting k-th and ($k+1$)-th vertices in π_{st}

3: **if** $i \neq j$ and $c\left(e_{[i,j]}\right) + c\left(e_{[i+1,j+1]}\right) < c\left(e_{[i,i+1]}\right) + c\left(e_{[j,j+1]}\right)$ **then**

4: **if** $i < j$ **then**

$\pi_{st} \leftarrow \langle v_s, \cdots, v_i, v_j, \cdots, v_{i+1}, v_{j+1}, \cdots, v_t \rangle$

else

$\pi_{st} \leftarrow \langle v_s, \cdots, v_j, v_i, \cdots, v_{j+1}, v_{i+1}, \cdots, v_t \rangle$

end if

$c(\pi_{st}) \leftarrow c(\pi_{st}) - c\left(e_{[i,i+1]}\right) - c\left(e_{[j,j+1]}\right) + c\left(e_{[i,j]}\right) + c\left(e_{[i+1,j+1]}\right)$

end if

end for

end for

Here, the final π_{st}^{*} is obtained corresponds to the unique Pareto-optimal complete path in the actual graph $G = (V, E, c)$.

11.4 COMPUTATIONAL EXPERIMENTS AND DISCUSSIONS

To test the proposed solution scheme, computational experiments are done on a real-world satellite image. The experiments are conducted on a machine with a 2.20 GHz processor and 8 GB RAM. The path searching algorithms in the solution scheme are coded in C++ while the *k*-Means clustering process is conducted on Wolfram Mathematica 11.2.0.0 using the ClusteringComponents[] command.

We extracted the aerial satellite image from part of Kedah State, Malaysia, with the approximate size of 15 km × 10 km from Google Earth Pro and the US Geological Survey website. Figure 11.8 shows the original satellite image where a region of 10 km × 10 km is extracted from within. The satellite images are rendered, and the terrain data is extracted using LandSerf 2.3. The images are downscaled into a grid map of dimension 200 × 200 and the attribute costs are estimated from the darkness level of the cells. The attribute costs are then normalised into a uniform range of $[0,100]$. Figure 11.9 shows the grayscale profile rendered for each of the three attribute costs corresponding to the region extracted from the original satellite image.

The computational experiments consist of four sets of tests, with each test containing 20 randomly assigned checkpoints with minimal distance of 2 cells apart. The list of coordinates of checkpoints for each problem

FIGURE 11.8 Original satellite image of Kedah State, Malaysia, before editing (approximated dimension 15 km × 10 km).

FIGURE 11.9 Grayscale image profile for three attributes (from left to right: traverse time, risk level, jamming level).

set is presented in Table 11.1. For each problem set, two subproblems with Λ being partitioned into 5 and 6 clusters, respectively, are formed. Each problem set is tested using the solution method as described in section 11.3, with and without the enhancement global 2-opt algorithm. Results obtained from the experiments are then compared.

In these experiments, the maximal computation time allowed is 10 minutes according to actual military practices Bae et al. (2015). For each

TABLE 11.1 List of Checkpoints for Every Problem Set

Problem Set	List of Checkpoints (Coordinate of Cells in Grid)
1	Λ = {(39,19),(27,99), (29,78),(40,128),(55,83),(64,96),(71,109),(76,96),(86,58), (93,99),(112,128),(134,77),(153,81),(172,129),(112,14),(135,36),(164,49), (168,50), (186,1),(80,74)}with v_s = (39,19) and v_t = (80,74)
2	Λ = {(29,34),(193,57),(31,71),(181,34),(78,99),(165,59),(83,63),(160,95), (89,38), (192,156),(61,10),(110,88),(49,62),(178,97),(21,96),(38,106),(23,3), (7,133), (186,80),(167,70)}with v_s = (29,34) and v_t = (167,70)
3	Λ = {(50,64),(59,75),(59,3),(40,79),(72,58),(123,14),(69,64),(90,56),(186,101), (179,122),(180,106),(199,20),(198,92),(185,118),(142,54),(101,116), (45,140),(64,152),(72,141),(129,113)}with v_s = (50,64) and v_t = (129,113)
4	Λ = {(128,55),(11,24),(100,41),(60,63),(76,3),(175,56),(31,119),(44,11),(89,94), (174,104),(3,26),(184,58),(101,22),(188,104),(42,31),(4,12),(195,116), (98, 135), (102,52),(96,56)}with v_s = (128;55) and v_t = (96;56)

problem instance, various combinations of weight **w** that reflects the priority given to each attribute are used:

- (0.9, 0.05, 0.05), (0.05, 0.9, 0.05) and (0.05, 0.05, 0.9), which indicate high priority given to one attribute only.
- (0.45, 0.45, 0.1), (0.45, 0.1, 0.45) and (0.1, 0.45, 0.45), which indicate equally high priorities are given to two attributes.
- (0.33, 0.33, 0.34) indicates equal priorities given to all attributes.
- Arbitrarily assigned weights (0.7, 0.1, 0.2), (0.2, 0.7, 0.1) and (0.1, 0.2, 0.7).

The computation time and the solution quality corresponding to different **w** are recorded. For each problem instance, the ideal cost $c^* = \left(c_1^*, c_2^*, c_3^*\right)$ is obtained by solving the lexicographically optimal complete path individually. To measure the closeness of a solution path from the ideal state, we calculate the L_2 distance between the solution point from the ideal point in the solution space. The computation times for each problem instance (not including global 2-opt improvement) are tabulated in Table 11.2. The L_2 distance of the solution path cost, $c\left(\pi_{st}^*\right)$ from the ideal cost c^* are tabulated in Table 11.3. Also, the information regarding the clusters is presented in Table 11.4.

From Table 11.2, we can see that our solution scheme solved every problem set within 80 seconds, with the longest time taken being 79.38 seconds. Notice that when the checkpoint set $|\Lambda|$ is partitioned into fewer clusters,

TABLE 11.2 Computation Time for Experiments (Without Global 2-opt)

	Computation Time for Different Problem Setting (sec.)							
	Set 1		Set 2		Set 3		Set 4	
Weight, w	$p=5$	$p=6$	$p=5$	$p=6$	$p=5$	$p=6$	$p=5$	$p=6$
(0.9, 0.05, 0.05)	34.38	35.41	52.50	65.07	54.32	57.44	48.54	40.81
(0.05, 0.9, 0.05)	48.40	56.43	65.31	79.38	65.46	70.76	47.72	58.42
(0.05, 0.05, 0.9)	39.97	47.97	57.73	73.27	52.78	64.07	54.81	48.11
(0.7, 0.1, 0.2)	32.33	33.72	49.98	62.70	51.50	53.33	45.21	38.65
(0.2, 0.7, 0.1)	41.39	46.92	58.96	75.97	59.05	64.52	45.66	50.97
(0.1, 0.2, 0.7)	40.28	46.96	57.29	73.72	53.51	62.20	53.16	49.41
(0.45, 0.45, 0.1)	38.01	41.67	56.72	75.33	59.62	63.03	47.65	47.12
(0.45, 0.1, 0.45)	35.54	37.34	51.61	65.77	51.86	54.72	48.48	42.34
(0.1, 0.45, 0.45)	41.16	48.07	59.59	78.80	57.14	64.22	51.58	52.78
(0.33, 0.33, 0.34)	37.75	41.46	55.69	73.97	53.96	59.73	48.67	47.86

TABLE 11.3 L_2 Distance of Solution Path Costs from Ideal Cost (Without Global 2-opt)

	L_2 Distance							
	Set 1		Set 2		Set 3		Set 4	
Weight, w	$p=5$	$p=6$	$p=5$	$p=6$	$p=5$	$p=6$	$p=5$	$p=6$
(0.9, 0.05, 0.05)	36,057	29,258	15,793	26,683	23,428	29,182	52,226	15,858
(0.05, 0.9, 0.05)	33,375	30,734	21,564	33,907	34,371	35,966	59,556	17,762
(0.05, 0.05, 0.9)	40,477	32,560	18,845	41,277	18,652	29,307	60,814	20,238
(0.7, 0.1, 0.2)	34,872	27,461	11,497	21,833	19,604	27,288	48,801	12,073
(0.2, 0.7, 0.1)	31,470	26,208	9515	29,828	28,480	31,109	55,981	13,837
(0.1, 0.2, 0.7)	35,948	29,969	11,796	22,679	13,494	26,507	49,890	14,199
(0.45, 0.45, 0.1)	31,790	24,761	4984	23,900	20,740	21,033	45,740	8601
(0.45, 0.1, 0.45)	34,389	28,515	10,643	24,008	19,205	21,200	46,286	7817
(0.1, 0.45, 0.45)	28,285	24,381	9892	25,829	18,055	21,968	48,887	9655
(0.33, 0.33, 0.34)	30,497	23,102	5775	21,032	14,161	18,474	44,403	7122

TABLE 11.4 Information about the Clusters in Each Problem Set

	Set 1		Set 2		Set 3		Set 4	
	$p=5$	$p=6$	$p=5$	$p=6$	$p=5$	$p=6$	$p=5$	$p=6$
Cluster radius:								
Min	52.50	60.42	78.72	36.80	60.93	35.51	76.54	43.66
Max	111.52	75.13	122.50	122.50	105.42	78.34	157.72	81.41
Average	79.62	67.95	95.36	75.12	90.10	62.75	108.24	60.95
Intercluster distance:								
Min	17.09	15.81	11.18	11.18	14.21	14.21	7.21	7.21
Max	84.15	111.31	97.35	138.28	92.54	136.62	68.47	135.33
Average	43.56	49.90	44.58	59.71	50.49	63.41	35.01	53.58

the computation time is less. This is because there are fewer intracluster TSPPs that need to be computed in problem sets with fewer clusters.

From both Table 11.3 and Table 11.4, we can see that when the average intercluster edge distance gets larger, the complete path π^*_{st} is further away from the ideal cost. This is possibly due to the following reasons:

1. The procedure in the solution scheme begins by first determining the shortest cluster visiting sequence together with its entering and leaving checkpoints for respective clusters. This might cause the optimality of the intercluster path to be sacrificed since the endpoints are predetermined.
2. In some occasions the checkpoints are surrounded with high attribute cost regions. This forces the path to take a longer tour to avoid the high-cost region. However, taking a detour to minimise travel costs for one attribute often sacrifices the cumulative travel costs of other attributes.
3. When the path visits a cluster with a large radius, it needs to travel a longer journey to visit every checkpoint in that cluster before moving on to visit the next cluster. This causes the solution path to accumulate cost at a much faster pace.

The situation is observed in Problem Set 4. For Problem Set 4 with $p=5$, notice that the average cluster radius is 77% larger than the one with $p=6$. The larger cluster radius means that the solution path needs to traverse more cells to visit all checkpoints within the cluster before moving on to visit other clusters. We can conclude that the checkpoints need to be partitioned into clusters with smaller radii and be closer to one another to yield a good solution path.

11.4.1 Solution Improvements Using 2-Opt

To improve the quality of solution paths, global 2-opt is performed on all complete paths. The computation time for 2-opt is presented in Table 11.5 and the results are presented in Table 11.6.

From Table 11.5 and Table 11.6, it is observed that most of the solution costs show improvement after global 2-opt is implemented. However, some of these improvements take additional time, ranging from 3 to 7 minutes with a maximum of 436.14 seconds. The reason global 2-opt requires such a large amount of evaluation time is because the 2-opt process needs to

TABLE 11.5 Computation Time for Experiments with Global 2-opt

	Computation Time for Different Problem Setting (sec.)							
	Set 1		Set 2		Set 3		Set 4	
Weight, *w*	*P* = 5	*P* = 6	*P* = 5	*P* = 6	*P* = 5	*P* = 6	*P* = 5	*P* = 6
(0.9, 0.05, 0.05)	193.20	188.54	201.38	342.66	285.08	304.71	253.85	261.07
(0.05, 0.9, 0.05)	340.07	344.78	369.65	436.14	300.10	332.83	329.78	362.94
(0.05, 0.05, 0.9)	270.49	293.96	371.63	399.12	221.87	241.67	319.59	323.39
(0.7, 0.1, 0.2)	221.86	218.72	272.64	370.82	249.47	272.18	267.04	281.28
(0.2, 0.7, 0.1)	342.55	320.06	355.14	424.35	273.81	305.75	322.80	343.14
(0.1, 0.2, 0.7)	277.96	301.52	368.67	396.83	226.25	255.58	297.20	323.35
(0.45, 0.45, 0.1)	307.66	296.69	339.74	420.20	290.94	319.55	323.48	349.56
(0.45, 0.1, 0.45)	260.35	267.50	321.46	384.30	239.84	271.71	300.19	322.20
(0.1, 0.45, 0.45)	301.48	305.96	357.78	410.53	259.15	292.17	308.27	334.32
(0.33, 0.33, 0.34)	291.23	282.89	338.57	381.60	275.93	304.82	314.30	344.01

TABLE 11.6 L_2 Distance of Solution Path Costs from Ideal Cost after Performing Global 2-opt

	L_2 Distance							
	Set 1		Set 2		Set 3		Set 4	
Weight, *w*	*P* = 5	*P* = 6	*P* = 5	*P* = 6	*P* = 5	*P* = 6	*P* = 5	*P* = 6
(0.9, 0.05, 0.05)	35,381	18,836	15,793	17,966	13,897	20,916	16,455	15,858
(0.05, 0.9, 0.05)	11,795	14,069	21,564	32,572	34,990	34,990	18,175	21,695
(0.05, 0.05, 0.9)	34,724	17,276	18,845	33,558	13,420	13,678	25,176	25,176
(0.7, 0.1, 0.2)	34,203	11,744	11,497	14,520	10,386	10,386	12,669	12,073
(0.2, 0.7, 0.1)	22,121	10,935	9515	22,405	29,390	29,390	28,059	14,127
(0.1, 0.2, 0.7)	24,830	10,292	11,796	17,684	10,295	11,571	16,391	16,391
(0.45, 0.45, 0.1)	31,259	9471	4984	14,456	23,766	23,766	9942	8601
(0.45, 0.1, 0.45)	33,501	11,512	10,643	15,474	9728	9728	7817	7817
(0.1, 0.45, 0.45)	22,386	11,081	9892	19,751	18,844	18,844	11,024	11,024
(0.33, 0.33, 0.34)	22,481	7837	5775	12,877	14,836	14,836	7550	7122

examine all pairs of edges where some of the edge pairs have not yet been determined.

On the other hand, for Problem Set 2 with $p = 5$, and some subproblems from Problem Set 4 with $p = 6$, no improvements are made after conducting global 2-opt. This is because the initial results obtained from these problems are already close to the ideal state. As a result, there is no more improvement to be made.

11.5 CONCLUSION

In this chapter, an extended GPP problem is modelled as a two-level clustered TSPP on a grid. This problem showcases the integration of three different aspects: grid-based path planning, multicriteria optimisation and TSPP. A two-phase solution scheme is developed to search for Pareto-optimal complete solution paths that traverse through all given checkpoints. Computational experiments show that the quality of the complete paths produced are highly influenced by the way the checkpoints are clustered, as well as the weights assigned on each cost attribute. To produce a solution path that is close to the ideal state, the clusters have to be close enough to each other, as well as having a relatively small cluster radius.

This research can be extended in several aspects. For example, one may consider different clustering techniques used to partition the checkpoints. Other than that, the relationship between the number of clusters and the solution quality can be explored to develop a better solution system for the problem. It is also possible to consider real-time cost functions instead of estimated terrain costs.

11.6 ACKNOWLEDGEMENT

The research work is supported by Universiti Sains Malaysia RUI research grant 1001/PMATHS/8011017.

REFERENCES

Bae, K. Y., Kim, Y. D., & Han, J. H. (2015). Finding a risk-constrained shortest path for an unmanned combat vehicle. *Computers & Industrial Engineering, 80*, 245–253. https://doi.org/10.1016/j.cie.2014.12.016

Baras, N., & Dasygenis, M. (2023). UGV Coverage path planning: An energy-efficient approach through turn reduction. *Electronics, 12*(13), 2959. https://doi.org/10.3390/electronics12132959

Benson, H. P., & Sun, E. (2002). A weight set decomposition algorithm for finding all efficient extreme points in the outcome set of a multiple objective linear program. *European Journal of Operational Research, 139*(1), 26–41. https://doi.org/10.1016/S0377-2217(01)00153-9

Ehrgott, M. (2008). Multiobjective optimization. *AI Magazine, 29*(4), 47–57. https://doi.org/10.1609/aimag.v29i4.2198

Giesbrecht, J. (2004). *Global path planning for unmanned ground vehicles*. Defence Research and Development Suffield.

Han, D. H., Kim, Y. D., & Lee, J. Y. (2014). Multiple-criterion shortest path algorithms for global path planning of unmanned combat vehicles. *Computers & Industrial Engineering, 71*, 57–69. https://doi.org/10.1016/j.cie.2014.02.013

Lu, Y., Hao, J. K., & Wu, Q. (2022). Solving the clustered traveling salesman problem via traveling salesman problem methods. *PeerJ Computer Science, 8*, e972. https://doi.org/10.7717/peerj-cs.972

Luc, D. T. (2008). The pareto optimality. *Pareto Optimality, Game Theory and Equilibria*, 481–515.

Marler, R. T., & Arora, J. S. (2010). The weighted sum method for multi-objective optimization: New insights. *Structural and Multidisciplinary Optimization, 41*(6), 853–862. https://doi.org/10.1007/s00158-009-0460-7

Mestria, M. (2018). New hybrid heuristic algorithm for the clustered traveling salesman problem. *Computers & Industrial Engineering, 116*, 1–12. https://doi.org/10.1016/j.cie.2017.12.018

Przybylski, A., Gandibleux, X., & Ehrgott, M. (2010). A recursive algorithm for finding all nondominated extreme points in the outcome set of a multiobjective integer programme. *INFORMS Journal on Computing, 22*(3), 371–386. https://doi.org/10.1287/ijoc.1090.0342

Saw, V., Rahman, A., & Ong, W. E. (2017a). Shortest path problem on a grid network with unordered intermediate points. *Journal of Physics: Conference Series, 893*(1), 012066. IOP Publishing. https://doi.org/10.1088/1742-6596/893/1/012066

Saw, V., Rahman, A., & Ong, W. E. (2017b). A weight assignment approach for solving multicriteria global path planning of unmanned combat vehicles. In *Proceedings of the 8th multidisciplinary international conference on scheduling: Theory and applications* (MISTA 2017, pp. 43–58). https://www.inprincipo.nl/file/upload/doc/mista2017.pdf

Sheng, W., Xi, N., Song, M., & Chen, Y. (2005). Robot path planning for dimensional measurement in automotive manufacturing. *Journal of Manufacturing Science and Engineering, 127*(2), 420–428. https://doi.org/10.1115/1.1870013

Thoresen, M., Nielsen, N. H., Mathiassen, K., & Pettersen, K. Y. (2021). Path planning for UGVs based on traversability hybrid a*. *IEEE Robotics and Automation Letters, 6*(2), 1216–1223. https://doi.org/10.1109/LRA.2021.3056028

U.S. Geological Survey. (n.d.). *Earth explorer.* https://earthexplorer.usgs.gov/

Wei, M., Wang, S., Zheng, J., & Chen, D. (2018). UGV navigation optimization aided by reinforcement learning-based path tracking. *IEEE Access, 6*, 57814–57825. https://doi.org/10.1109/ACCESS.2018.2872751

CHAPTER 12

Energy and Exergy Analyses of Hybrid Solar Drier for Seaweed Drying

Raja Aqib Shamim, Majid Khan
Majahar Ali and Ahmad Fudholi

12.1 INTRODUCTION: BACKGROUND AND DRIVING FORCES

Solar energy as an alternative energy source is becoming increasingly attractive. It is a clean energy using the world's most abundant permanent energy source (Singh, 2013). It also has the potential to meet a significant proportion of the world's energy needs. In Malaysia, solar drying systems have great potential for drying purposes, which is relevant to the local climate, with an amount of 4.21 to $5.56\,kWh/m^2$ daily solar radiation intensity. The solar drying system is one of the most attractive and promising applications of solar energy systems (Kant et al., 2016). The global demand for premium quality dry agricultural and marine products is exceptionally high. To meet this demand, the majority of these commodities must undergo a drying process to achieve the desired level of quality. Traditionally, this process involved drying most crops under the sun

DOI: 10.1201/9781032678573-12

(Jain & Tiwari, 2003; Rathore & Panwar, 2021). Drying remains one of the most important post-handling processes for agricultural produce. It can extend the shelf life of the harvested products, improve quality, improve the bargaining position of the farmer to maintain relatively constant prices of their products, reduce post-harvest losses and lower transportation costs since most of the water is taken out from the product during the drying process. Direct sun drying requires a large open space area and is completely dependent on the availability of sunshine and susceptible to contamination with foreign materials such as dust and litter (Ong, 1999). Furthermore, it is exposed to birds, insects and rodents. Hence, most agricultural produce that is intended to be stored must be dried first. Otherwise, insects and fungi, which thrive in moist conditions, render them unusable (Sopian et al., 2012; Fudholi et al., 2010; Ruslan et al., 2011).

Recently, many reports have been published on the drying kinetics of agricultural fruits and vegetables (Dissa et al., 2011; Doymaz, 2010; Amer et al., 2010). Thin-layer drying models also have been widely used for the analysis of the drying of various marine products (Duan et al., 2010; Dissa et al., 2010; Kilic, 2009). Although many mathematical models have been proposed to describe the drying process, a limited number of reported studies on mathematical models and drying kinetics of marine products such as seaweed are in the literature. Recently, seaweed has been widely used in the fabrication of food, medical industries and industry manufacture. Seaweed is a potential renewable energy source, and it can be converted into energy such as biofuel, oil and gas (Demirbas, 2010; Amin, 2009). Demirbas (2010) reported that seaweed can be used to produce bio-oil, biodiesel, ethanol, methane and hydrogen. By using thermochemical and biochemical processes, oil, gas, ethanol, biodiesel and bio-hydrogen can be produced (Amin, 2009). Ge et al. (2011) reported that brown seaweed (*Laminaria japonica*) can be used to produce alginate, iodine and mannitol. They reported that seaweed has excellent prospects as a potential feedstock for the production of bioethanol. Seaweed as an energy resource to produce biogas has been studied by Vergara-Fernández et al. (2008). Gupta et al. (2011) studied the effect of different temperatures on the drying kinetics and the phytochemical constituents of edible Irish brown seaweed (*Himanthalia elongate*). The study involved the modeling of the term of Fick's diffusion equation for the estimation of the diffusion coefficients. They concluded that the drying kinetics of seaweed can be accurately predicted using Newton's model, the logarithmic model and Henderson and Pabis's model. Fudholi et al. (2011a) reported the effects of air temperature and humidity on the drying kinetics of red seaweed.

The drying kinetics of red seaweed were studied using a solar drying system (Fudholi et al., 2011b), whereas a hot air chamber was used to determine the drying kinetics of brown seaweed (*Eucheuma cottonii*; Fudholi et al., 2012a). The present study was carried out to select the best mathematical model to illustrate the drying behavior of red seaweed using a solar drying system.

Thermodynamics plays an important role in enhancing the energy efficiency of industrial processes. The energy used in a system or process is significant and therefore represents an often reducible element of process cost. It is possible to identify the operation conditions in which potential savings can be made using an exergy analysis. Exergy is defined as the maximum amount of work that can be produced by a stream of matter, heat or work as it comes to equilibrium with a reference environment. Also, exergy analysis is a useful method to establish strategies for the design and operation of many industrial processes where the optimal use of energy is considered an important issue. This information is effective in determining the plant and the operation cost, the energy conservation, the fuel versatility and the pollutant. In recent years, exergy analysis has been widely used for the performance evaluation of thermal systems. In the drying process, the aim is to use the minimum amount of energy for maximum moisture removal for the desired final conditions of the products (Akbulut & Durmuş, 2010; Akpinar, 2010). Numerous studies have focused on exergy analyses of food drying. However, a thorough literature review for the present study reveals a significant gap: there is no information on energy and exergy analyses of solar drying systems specifically for seaweed. Previous research on experimental solar drying has not applied energy and exergy analysis methods, leaving a lack of data on the performance of solar drying for seaweed. The potential for improving solar drying systems is also scarcely addressed in the existing literature. The main objective of this chapter is to analyse the energy and exergy of solar drying systems for red seaweed. Therefore, this chapter, like other studies, concentrates on the performance, drying model and exergy analyses of solar drying systems for red seaweed.

12.2 MATERIAL AND METHODS

12.2.1 Material

The red seaweed used in this study was obtained from Kedah, Malaysia. The initial moisture content of red seaweed was determined by measuring its initial and final weight using the hot air chamber at $120°C$ until a constant weight was obtained (Meziane, 2011). The average initial moisture content of red seaweed was obtained to be 92.68% wet basis.

12.2.2 Experimental Apparatus

A solar drying system was installed at the Green Energy Technology Innovation Park, Universiti Kebangsaan Malaysia (UKM), Malaysia, in 2010. The drier is classified as a forced convection indirect type. A schematic diagram of the experimental solar drying system is shown in Figure 12.1. It consists of an auxiliary heater, blower, drying chamber and double-pass solar collector.

The collector width and length are $1.2\,m$ and $4.8\,m$, respectively. The solar collector array consists of four solar collectors as shown in Figure 12.2. The upper channel depth is $3.5\,cm$ and the lower depth is $7\,cm$. The bottom and sides of the collector have been insulated with $2.5\,cm$ thick fiberglass to minimise heat losses. The cross-section of a double-pass solar collector

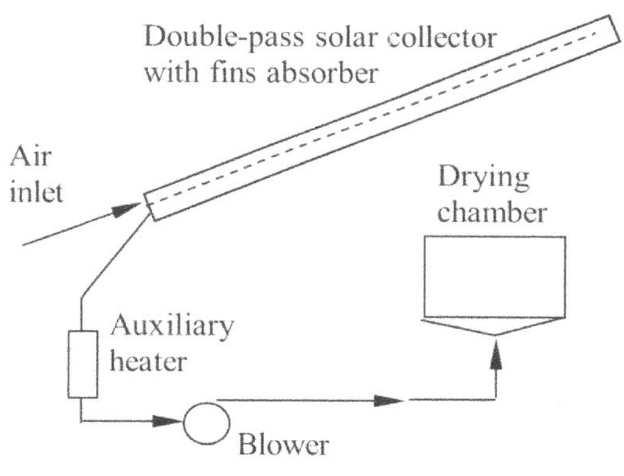

FIGURE 12.1 Schematic diagram of solar drying system.

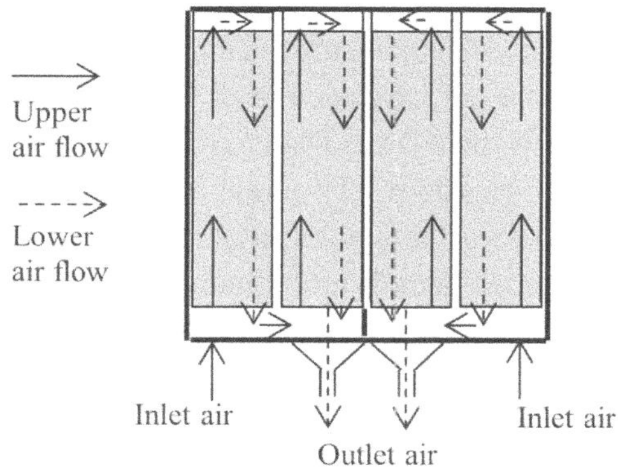

FIGURE 12.2 The collectors of solar drying system.

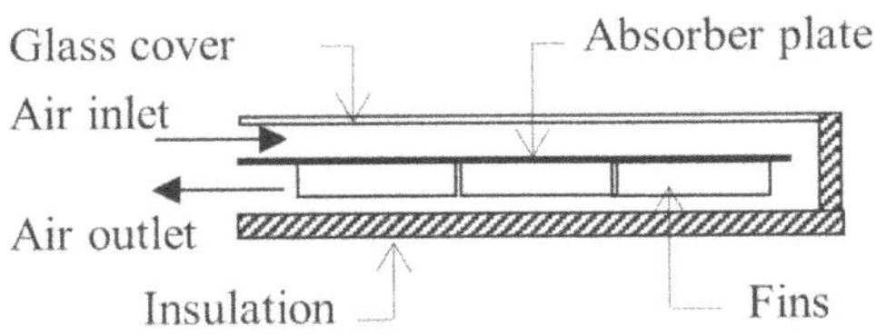

FIGURE 12.3 The schematic of a double-pass solar collector with finned absorber.

with a finned absorber is shown in Figure 12.3. The collector consists of the glass cover, the insulation and the black-painted aluminum absorber. The size of the collector is $1.2m$ wide and $4.8m$ long. In this type of collector, the air initially enters through the first channel formed by the glass covering the absorber plate and then through the second channel formed by the back plate and the finned absorber plate. The size of the chamber is $2.4m$ in length, $1m$ in width and $0.6m$ in height.

12.2.3 Experimental Procedure and Uncertainties

The experiment was done between 08:00 and 18:00. During the drying experiments, the weather was sunny and no rain appeared. Broken seaweed was loaded over the trays of a drying chamber. The data measured were air temperature (ambient temperature, air temperature inlet and outlet of the collector), radiation intensity and air velocity, also measured the air temperature before it enters the drier chamber, the temperature inside the drier chamber and the temperature of the air out of the drier chamber. The air temperature was measured by a T-type thermocouple with an accuracy of $0.018°C$. Relative humidity sensors were installed in the inlet, middle and outlet sections of the drying chamber. An airflow DTA 4000 anemometer to determine the airflow velocity in the solar collector was used. An LI-200 pyranometer with 1% accuracy was used. During the drying process, the temperature and relative humidity in the solar drier were recorded at 1-minute intervals during the experiments with the ADAM Data Acquisition System connected to a computer. Data were averaged for 30 minutes before analysis.

A drying experiment was done for a large amount of red seaweed: $40\,kg$ of red seaweed was divided equally and then placed on eight trays as shown in Figure 12.4. To determine the moisture loss of drying red seaweed during experiments, seaweed was placed at the center of the drying chamber in a small tray. The moisture loss of red seaweed was determined through a Camry R9364 digital electronic balance having an accuracy of $0.01\,g$ on the top center of the drying chamber.

FIGURE 12.4 Photograph of seaweed in drying chamber.

TABLE 12.1 Uncertainties during the Measurements of the Parameters

Parameters	Unit	Uncertainty Comment
Ambient air temperature	°C	±0.15
Collector inlet temperature	°C	±0.24
Collector outlet temperature	°C	±0.24
Chamber temperature	°C	±0.24
Solar intensity	W/m^2	±0.4
Air velocity	m/s	±0.2
Relative humidity	%	±0.16

During the measurements of the parameters, the uncertainties that occurred are shown in Table 12.1. The estimation of uncertainties was calculated by El-Sebaii et al. (2011):

$$X_R = \left[(x_1)^2 + (x_2)^2 + \ldots (x_n)^2\right]^{1/2} \tag{12.1}$$

where

X_R = uncertainty in the result,
$x_1, x_2 \ldots x_n$ = uncertainty in the independent variables.

12.2.4 Mathematical Modelling of Drying Curves

The moisture content was expressed as a percent wet basis. The experimental drying data for red seaweed were fitted to the thin layer drying models in Table 12.2, the moisture ratio (*MR*) can be calculated as (Fruit, 2012)

$$MR = \frac{M - M_e}{M_0 - M_e} \tag{12.2}$$

where

M_e = equilibrium moisture content,
M_0 = initial moisture content.

The moisture content of materials (M) can be calculated by two methods based on either wet or dry basis using the following equations.

The moisture content wet basis is

$$M = \frac{w(t) - d}{w} \times 100\% \tag{12.3}$$

The moisture content dry basis is

$$X = \frac{w(t) - d}{d} \tag{12.4}$$

where

$w(t)$ = mass of wet materials at instant t,
d = mass of dry materials.

The values of the coefficient of determination (R^2), mean bias error (MBE) and root mean square error (RMSE) were used to determine the quality of the drying model. The highest R^2 values, the values of MBE and

TABLE 12.2 Several Models of Drying

No.	Model Name	Model
1	Newton	$MR = exp(-kt)$
2	Page	$MR = exp(-kt^n)$
3	Henderson and Pabis	$MR = a\ exp(-kt)$

the lowest RMSE were selected to estimate which drying curve is the best (Othman et al., 2012; Tunde-Akintunde, 2011).

$$MBE = \frac{1}{N}\sum_{i=1}^{N}\left(MR_{pre,i} - MR_{\exp,i}\right)^2 \tag{12.5}$$

$$RMSE = \left[\frac{1}{N}\sum_{i=1}^{N}\left(MR_{pre,i} - MR_{\exp,i}\right)^2\right]^{\frac{1}{2}} \tag{12.6}$$

where

N = number of observations,
$MR_{pre,i}$ = predicted moisture ratio,
$MR_{exp,i}$ = experimental moisture ratio.

12.2.5 Energy Analysis

The specific energy consumption (SEC) of the solar drying system was obtained using Eq. (12.7), as reported by Fudholi et al. (2013)

$$SEC = \frac{P_t}{W} \tag{12.7}$$

where

SEC = specific energy consumption (kWh/kg),
W = mass of water evaporated from the product (kg),
P_t = total energy input to the drier (kWh).

The mass of water removed (W) from a wet product can be calculated by

$$W = \frac{m_o\left(M_i - M_f\right)}{100 - M_f} \tag{12.8}$$

where

m_o = initial total crop mass (kg),
M_i = initial moisture content fraction on a wet basis,
M_f = the final moisture content fraction on a wet basis.

The thermal efficiency of the solar collector η_c was estimated by the following equation

$$\eta_c = \frac{mC(T_o - T_i)}{A_c S} \times 100\% \tag{12.9}$$

where

m = mass flow rate (kg/s),
C = specific heat of the air ($Jkg^{-1}°C^{-1}$),
A_c = collector area (m^2),
T_i = inlet air temperature (°C),
T_o = outlet air temperature (°C),
S = solar radiation intensity (W/m^2).

System drying efficiency was defined as the ratio of the energy required to evaporate from the moisture to the heat supplied to the drier. The system efficiency (η) for forced convection solar driers needs to take into account the energy consumed by the fan/blower, and it can be calculated as (Fudholi, 2012b)

$$\eta = \frac{WL}{SA_c + P_f} \tag{12.10}$$

where

W = weight of water evaporated from the product (kg),
L = latent heat of vaporisation of water at exit air temperature (J/kg),
P_f = power of fan.

Pick-up efficiency (η) determines the efficiency of moisture removal by the drying air from the product, which can be calculated as

$$\eta = \frac{h_0 - h_i}{h_{as} - h_i} = \frac{W}{\nu\rho t(h_{as} - h_i)} \tag{12.11}$$

where

h_o = absolute humidity of air leaving the drying chamber (%),
h_i = absolute humidity of air entering the drying chamber (%),
h_{as} = absolute humidity of the air entering the drier at the point of adiabatic saturation (%),
ν = volumetric airflow (m^3/s),
ρ = density of air (kg/m^3),
t = drying time (s).

12.2.6 Exergy Analysis

The exergy values are calculated using characteristics of the working medium from a first law of energy balance. For this purpose, the general form of the exergy equation applicable to a steady flow system may be expressed as (Akbulut & Durmuş, 2010)

$$Ex = \dot{m}_{da} C_{pda} \left[\left(T - T_a \right) - T_a \ln \frac{T}{T_a} \right] \quad (12.12)$$

where

Ex = exergy,
$\dot{m}_{da}$ = mass of the drying ambient,
C_{pda} = pressure coefficient of drying ambient,
T_a = the temperature of the ambient.

For exergy inflow of drying chamber:

$$Ex_{dci} = \dot{m}_{da} C_{pda} \left[\left(T_{dci} - T_a \right) - T_a \ln \frac{T_{dci}}{T_a} \right] \quad (12.13)$$

where

Ex_{dci} = exergy for drying inlet chamber,
T_{dci} = the temperature of the drying chamber inlet.

For exergy outflow of the drying chamber:

$$Ex_{dco} = \dot{m}_{da} C_{pda} \left[\left(T_{dco} - T_a \right) - T_a \ln \frac{T_{dco}}{T_a} \right] \quad (12.14)$$

where

Ex_{dco} = exergy for drying outlet chamber,
T_{dco} = the temperature of the drying chamber outlet,

However, during the solar drying process, the exergy losses Ex_{loss} are determined using the following equation:

$$Ex_{loss} = Ex_{dci} - Ex_{dco} \quad (12.15)$$

The exergy efficiency can be defined as the ratio of energy used (investment) in the drying of the product to the exergy of the drying air supplied to the system. However, it is explained as the ratio of exergy outflow to exergy inflow for the drying chamber. Considering this definition, the exergy efficiencies of the drying chamber can be determined. Thus, the general form of exergy efficiency η_{Ex} is written as (Akbulut & Durmuş, 2010; Akpinar, 2010)

$$\eta_{Ex} = \frac{Ex_{dco}}{Ex_{dci}} = 1 - \frac{Ex_{loss}}{Ex_{dci}} \tag{12.16}$$

Soares et al. (2007) have proposed that maximum improvement in the exergy efficiency for a system or process is achieved when the exergy loss (Ex_{loss}) is minimised. They suggested that it is useful to employ the concept of an exergy "improvement potential" when analysing different processes or sectors of the economy. The improvement potential (*IP*) of a system or process is given by

$$IP = \left(1 - \eta_{Ex}\right) Ex_{loss} \tag{12.17}$$

12.3 RESULTS AND DISCUSSION

The variations of the solar radiation (S), collector inlet air temperature (T_i), collector outlet air temperature (T_o), drying chamber air temperature (T_c) and average relative humidity of the drying chamber (H_c) over the 2 days of experiments are shown in Figure 12.5. From these figures, the daily mean of the drying chamber air temperature, relative humidity of the drying chamber and solar radiation range from about $35-60°C$, $32-63\%$ and $83-920\,W/m^2$, respectively. During the 2 days of drying, the daily averages of relative humidity at the drying chamber were obtained as 43 and 44%.

With about $40\,kg$ of red seaweed in the chamber, drying of seaweed takes 2 days to reduce the initial moisture content of 94.6% to 10%, equivalent to $40\,kg$ to $4.44\,kg$. During the 2 days of drying, the daily averages of air temperature at the drying chamber were 45 and $50°C$, and average solar radiations were 453 and $562\,W/m^2$ at a mass flow rate of $0.0536\,kg/s$, as shown in Figure 12.6. The efficiency of the collector varies from 26 to 80%, and the average efficiency of the collector was about 35%, as shown in Figure 12.6.

The kinetic curve of drying red seaweed at two conditions decreased moisture content wet basis of drying time, as shown in Figure 12.7. The drying curve can also show the profile change in moisture content (MR) versus

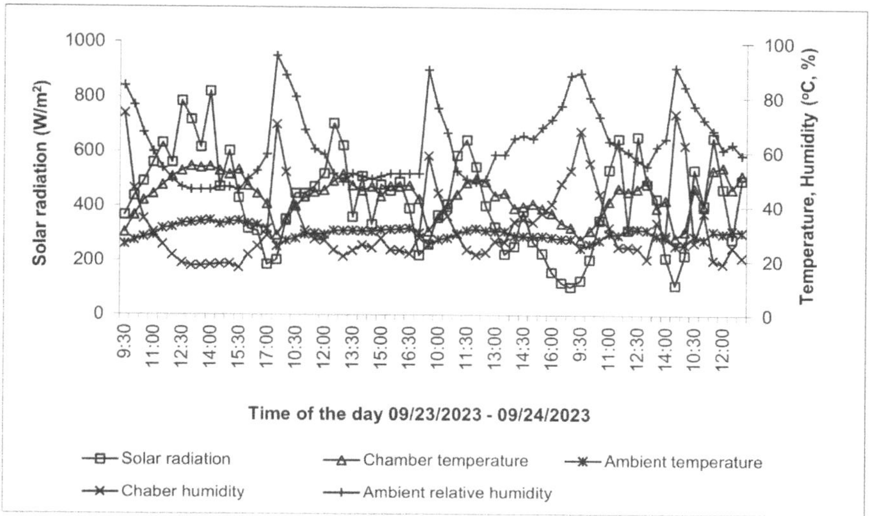

FIGURE 12.5 Temperatures (inlet, outlet and chamber), relative humidity of chamber and solar radiation.

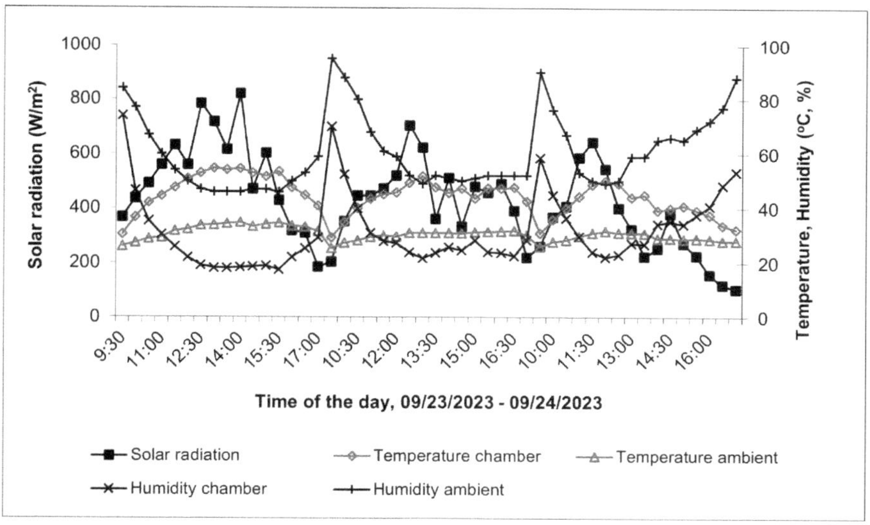

FIGURE 12.6 Efficiency of collector and solar radiation at $m = 0.0536$ kg/s for the 1st and 2nd drying day.

drying time (t), as shown in Figure 12.8. The drying kinetics of seaweed can be accurately predicted using Newton's model, the logarithmic model and Henderson and Pabis's model (Gupta et al., 2011). Azoubel et al. (2010) reported that Page's model improved the simulation in comparison with the

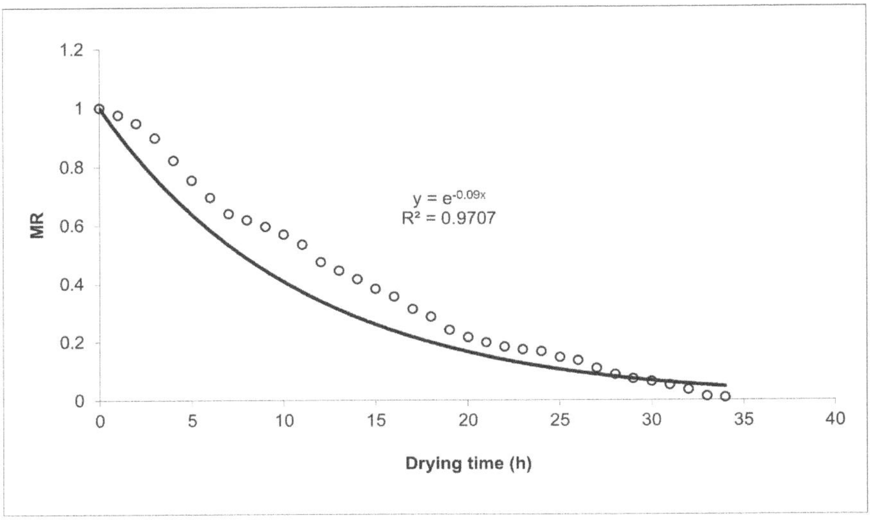

FIGURE 12.7 Drying curve: wet basis moisture content versus drying time.

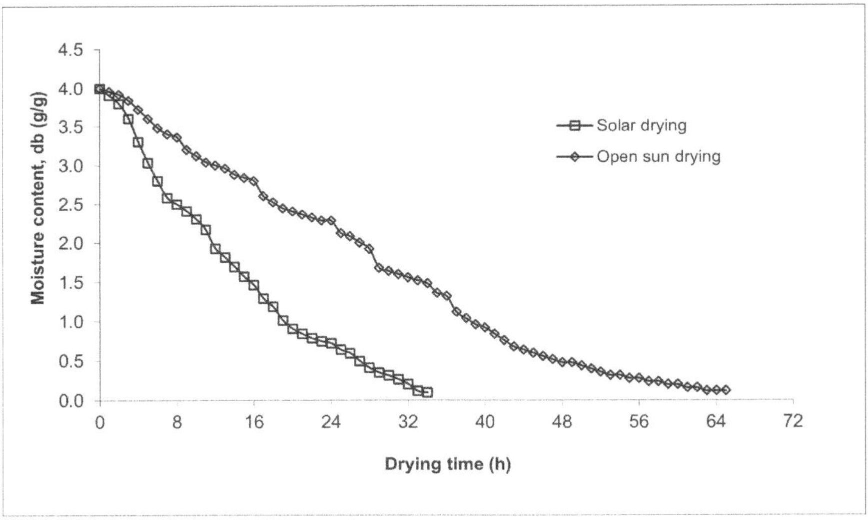

FIGURE 12.8 Drying curve: dry basis moisture content versus drying time.

results obtained using the diffusion model, having the best fit to the experimental data, with calculated average error ranging from 1.89% to 12.76% and R^2 values greater than 0.99. Page's model has shown a better fit than other models at accurately simulating the drying curves of chili pepper (Tunde-Akintunde, 2011), rapeseed (Han & Keum, 2011), green beans (Doymaz, 2005a), okra (Doymaz, 2005b) and kiwi (Simal et al., 2005), among others.

Page's equation can also be written to the equation

$$\ln(-\ln MR) = \ln k + n \ln t \tag{12.18}$$

Equation (12.18) is the relationship ln (-ln *MR*) versus *t*, which was the curve of the logarithmic equation, as shown in Figure 12.10. Henderson and Pabis's equation can also be written as

$$\ln MR = -kt + \ln a \tag{12.19}$$

From Eq. (12.19), a plot of ln *MR* versus drying time gives a straight line with intercept = $\ln a$, and slope $= k$. From the graph, *MR* versus $\ln t$, as shown in Figure 12.11, obtained the value $k = -0.109$ and the value of $\ln a = 0.4396$.

A set of experiments was conducted to develop a drying model to simulate the drying curves of red seaweed. Drying models were fitted with the experimental data of drying in the form of changes in moisture content versus drying time. In these drying models, changes in moisture content versus time are calculated using Microsoft Excel software. Fitting some of the drying models has been done with the experiment data, as shown in Figures 12.9 to 12.11. The values of R^2, MBE, RMSE and the parameters a, n and the constant k for the

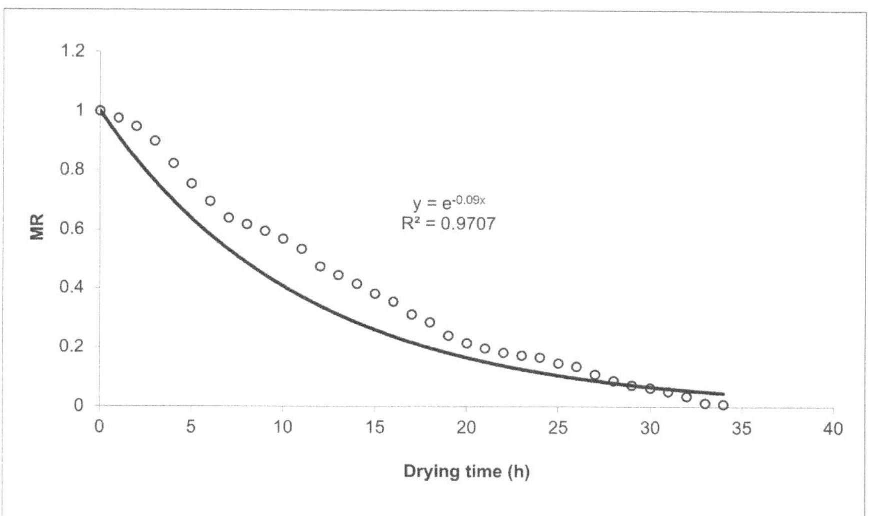

FIGURE 12.9 Plot of *MR* versus drying time (Newton's model).

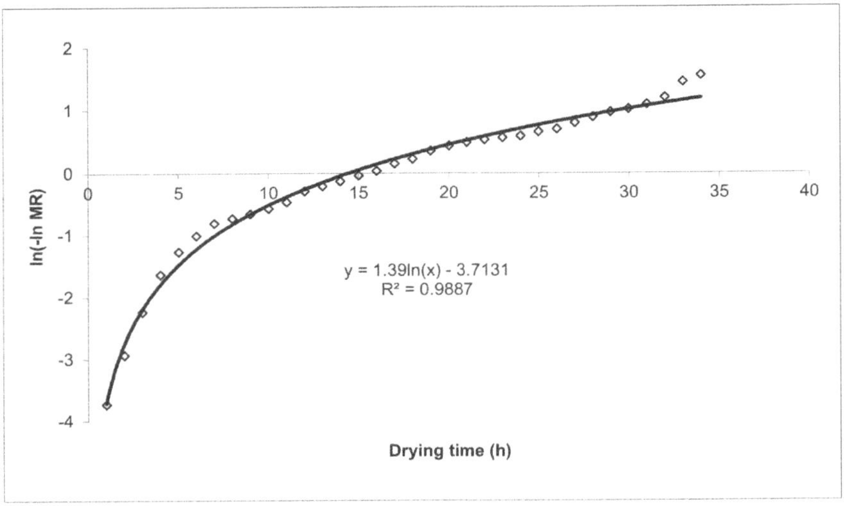

FIGURE 12.10 Plot of ln (–ln *MR*) versus drying time (Page's model).

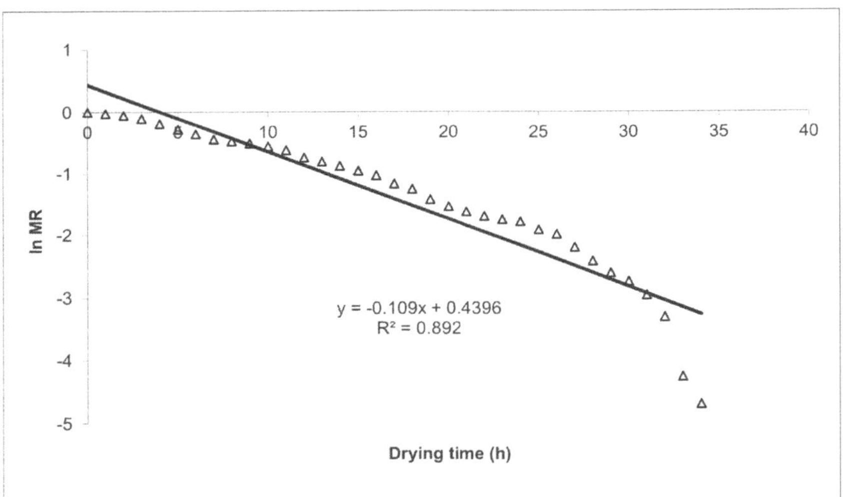

FIGURE 12.11 Plot of ln *MR* versus drying time (Henderson and Pabis model).

different models are listed in Table 12.3. The highest value of R^2 and lowest value of MBE and RMSE indicated the goodness of fit. Results presented in Table 12.3 show that Page's model has the highest value of R^2 (0.9676), as well as the lowest values of MBE (0.00023) and RMSE (0.01510), compared to Newton's model and Henderson and Pabis's model. Accordingly, Page's model was selected as the suitable model to represent the thin layer drying

TABLE 12.3 Constants Value Fitting of the Drying Models

Model	*a*	*k*	*n*	R^2	MBE	RMSE
Newton		0.4019		0.9033	0.00529	0.07274
Page		0.9442	0.6123	0.9676	0.00023	0.01510
Henderson and Pabis	0.5367	0.3327		0.9608	0.00024	0.01539

TABLE 12.4 Saturated Steam (Earle, 1983)

Temperature (°C)	Pressure (kPa)	Enthalpy (sat. vap.) (kJ/kg)	Latent Heat (kJ/kg)	Specific Volume (m^3/kg)
30	4.25	2556	2431	32.9
40	7.38	2574	2407	19.5
50	12.3	2592	2383	12.0
60	19.9	2610	2359	7.67
70	31.2	2627	2334	5.04

behaviour of red seaweed. This is in accordance with Fudholi et al. (2011a, 2011b, 2012a) that Page's model was shown to be a better fit for drying seaweed than other drying models. Regarding the drying behavior of lemongrass, Newton's model demonstrated a superior fit to the experimental data compared to other semi-theoretical models (Ibrahim et al., 2009).

The experimental results showed that solar drying without auxiliary heating of $40\,kg$ of dry red seaweed required about 10% water content within 15 hours (2 days of drying) to yield $4.44\,kg$ of dried red seaweed. Specific energy consumption of $2.62\,kWh/kg$ was obtained by using Eq. (12.7) for the specific energy consumption. However, the weight of water evaporated from the red seaweed obtained using Eq. (12.8) was $35.56\,kg$. Substituting $L = 2383\,kJ/kg$ $(662\,Wh/kg)$ for $T = 50°C$ as shown in Table 12.4, $t = 15$ hours and $S = 500\,W/m^2$ to Eq. (12.10) yielded a drying efficiency of 27.1%. By using Eq. (12.11) and a psychometric chart, the value of pick-up efficiency is 94.51%. During the solar drying process, the exergy efficiency was calculated by using Eq. (12.16), and it was in the range of 1% to 93% with an average of 30%, as shown in Figure 12.12. Exergy analysis can help minimise the use of natural resources by highlighting where the work potential of these resources is lost relative to the surrounding environment. By using Eq. (12.17), the values of improvement potential were found to be in the range of 0.3 and $630\,W$, with an average of $247\,W$ as shown in Figure 12.12. The summary of the experimental results and observations is given in Table 12.5.

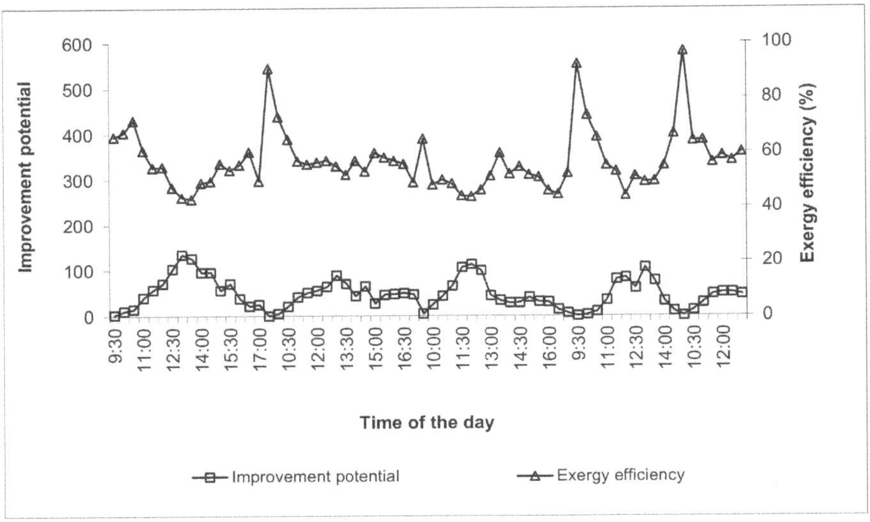

FIGURE 12.12 The variation of improvement potential and exergy efficiency as a function of time.

TABLE 12.5 Performance of Solar Drying for Red Seaweed

Parameters	Unit	Value
Initial weight (total)	kg	40
Final weight (total)	kg	4.44
Initial moisture content (wet basis)	%	90
Final moisture content (wet basis)	%	10
Mass flow rate	kg/s	0.05
Average solar radiation	W/m^2	500
Average ambient temperature	°C	30
Average drying chamber temperature	°C	48.6
Average drying chamber humidity	%	43
Drying time	h	15
Blower energy	kWh	0.45
Specific energy consumption	kWh/kg	2.62
Overall heat collection (thermal) efficiency	%	35
Overall drying efficiency, up to 10% wet basis	%	27
Pick-up efficiency, up to 10% wet basis	%	95
Overall exergy efficiency, up to 10% wet basis	%	30
Overall improvement potential	kW	0.247

12.4 CONCLUSION

A solar drying system was designed, constructed and tested for seaweed drying. Kinetic curves of seaweed drying are determined using the proposed thin-layer models. The non-linear regression procedure was used to fit three different drying models, Newton's model, Page's model and Henderson and Pabis's model. The models were compared with experimental data of red seaweed drying at the daily average air temperature of about $50°C$. The fit quality of the models was evaluated using the coefficient of determination (R^2), MBE and RMSE. The highest values of R^2 (0.9676), the lowest MBE (0.00023) and RMSE (0.001510) indicated that the Page model is the best mathematical model to describe the drying behavior of red seaweed.

The solar collector, drying system and pick-up efficiency rates of about 28%, 13% and 45%, respectively, at the average solar radiation of about $500\,W/m^2$ and airflow rate of $0.05\,kg/s$. A maximum and minimum collector efficiency of about 80% and 23%, respectively, were observed, and the drying temperature varied between $35°C$ and $60°C$, with an average of $48.6°C$. Specific energy consumption of $2.62\,kWh/kg$ was obtained. The values of exergy efficiency varied between 1 and 93%. The values of improvement potential were found to be in the range of 0.3 and $630\,W$, with an average of $247\,W$.

The study on the drying rate and quality of dried seaweed reveals a significant relationship between these two factors. The drying rate is a critical parameter that describes the drying characteristics of seaweed, influencing both the efficiency of the drying process and the preservation of its nutritional content. By optimising the drying rate, it is possible to maintain the seaweed's nutrients at optimal levels, thereby ensuring high-quality dried products. This is particularly important as improper drying can lead to nutrient degradation, affecting the overall quality of the seaweed.

Regarding the optimisation of hybrid solar drier design, future efforts should focus on expanding the collector area. This enhancement would allow for increased energy capture, thereby improving the overall efficiency of the drying system. By optimising the design in this manner, the system can achieve faster drying times and better energy utilisation, contributing to more sustainable and cost-effective drying processes.

For future research directions, it is recommended to continue exploring ways to enhance the energy efficiency and performance of hybrid solar driers. This includes not only expanding the collector area but also integrating advanced materials and technologies that can improve heat retention and distribution. Additionally, research should investigate the impact of different drying parameters on the quality of dried seaweed to develop more refined drying protocols that maximise nutrient retention while minimising energy consumption.

REFERENCES

Akbulut, A., & Durmuş, A. (2010). Energy and exergy analyses of thin layer drying of mulberry in a forced solar dryer. *Energy, 35*(4), 1754–1763. https://doi.org/10.1016/j.energy.2009.12.028

Akpinar, E. K. (2010). Drying of mint leaves in a solar dryer and under open sun: Modelling, performance analyses. *Energy Conversion and Management, 51*(12), 2407–2418. https://doi.org/10.1016/j.enconman.2010.05.005

Amer, B. M. A., Hossain, M. A., & Gottschalk, K. (2010). Design and performance evaluation of a new hybrid solar dryer for banana. *Energy Conversion and Management, 51*(4), 813–820. https://doi.org/10.1016/j.enconman.2009.11.016

Amin, S. (2009). Review on biofuel oil and gas production processes from microalgae. *Energy Conversion and Management, 50*(7), 1834–1840. https://doi.org/10.1016/j.enconman.2009.03.001

Azoubel, P. M., Baima, M. D. A. M., da Rocha Amorim, M., & Oliveira, S. S. B. (2010). Effect of ultrasound on banana cv Pacovan drying kinetics. *Journal of Food Engineering, 97*(2), 194–198. https://doi.org/10.1016/j.jfoodeng.2009.10.009

Demirbas, A. (2010). Use of algae as biofuel sources. *Energy Conversion and Management, 51*(12), 2738–2749. https://doi.org/10.1016/j.enconman.2010.06.010

Dissa, A. O., Bathiebo, D. J., Desmorieux, H., Coulibaly, O., & Koulidiati, J. (2011). Experimental characterisation and modelling of thin layer direct solar drying of Amelie and Brooks mangoes. *Energy, 36*(5), 2517–2527. https://doi.org/10.1016/j.energy.2011.01.044

Dissa, A. O., Desmorieux, H., Savadogo, P. W., Segda, B. G., & Koulidiati, J. (2010). Shrinkage, porosity and density behaviour during convective drying of spirulina. *Journal of Food Engineering, 97*(3), 410–418. https://doi.org/10.1016/j.jfoodeng.2009.10.036

Doymaz, İ. (2005a). Drying behaviour of green beans. *Journal of Food Engineering, 69*(2), 161–165. https://doi.org/10.1016/j.jfoodeng.2004.08.009

Doymaz, İ. (2005b). Drying characteristics and kinetics of okra. *Journal of Food Engineering, 69*(3), 275–279. https://doi.org/10.1016/j.jfoodeng.2004.08.019

Doymaz, İ. (2010). Effect of citric acid and blanching pre-treatments on drying and rehydration of Amasya red apples. *Food and Bioproducts Processing, 88*(2-3), 124–132. https://doi.org/10.1016/j.fbp.2009.09.003

Duan, X., Zhang, M., Mujumdar, A. S., & Wang, S. (2010). Microwave freeze drying of sea cucumber (*Stichopus japonicus*). *Journal of Food Engineering, 96*(4), 491–497. https://doi.org/10.1016/j.jfoodeng.2009.08.031

Earle, R. L. (1983). *Unit operations in food processing* (2nd ed., p. 189). Pergamon Press.

El-Sebaii, A. A., Aboul-Enein, S., Ramadan, M. R. I., Shalaby, S. M., & Moharram, B. M. (2011). Thermal performance investigation of double pass-finned plate solar air heater. *Applied Energy, 88*(5), 1727–1739. https://doi.org/10.1016/j.apenergy.2010.11.017

Fruit, O. D. (2012). Drying characteristics of the Borneo *Canarium odontophyllum* (Dabai) Fruit. *American Journal of Agricultural and Biological Sciences, 7*(3), 347–356. https://doi.org/10.3844/ajabssp.2012.347.356

Fudholi, A., Othman, M. Y., Ruslan, M. H., & Sopian, K. (2013). Drying of Malaysian *Capsicum annuum* L. (red chili) dried by open and solar drying. *International Journal of Photoenergy, 2013*(1), 167895. https://doi.org/10.1155/2013/167895

Fudholi, A., Othman, M. Y., Ruslan, M. H., Yahya, M., Zaharim, A., & Sopian, K. (2011a). The effects of drying air temperature and humidity on drying kinetics of seaweed. *Recent Research in Geography, Geology, Energy, Environment and Biomedicine*, 129–133. ISBN: 978-1-61804-022-0

Fudholi, A., Othman, M. Y., Ruslan, M. H., Yahya, M., Zaharim, A., & Sopian, K. (2011b). Design and testing of solar dryer for drying kinetics of seaweed in Malaysia. *Recent Research in Geography, Geology, Energy, Environment and Biomedicine*, 119–124. ISBN: 978-1-61804-022-0

Fudholi, A., Ruslan, M. H., Haw, L. C., Mat, S., Othman, M. Y., Zaharim, A., & Sopian, K. (2012a, January). Mathematical modeling of brown seaweed drying curves. In *Proceedings of the WSEAS international conference on applied mathematics in electrical and computer engineering* (pp. 207–211). Cambridge, USA: Harvard. ISBN: 978-1-61804-064-0. https://d1wqtxts1xzle7.cloudfront.net/53689263/MATHCC-32-libre.pdf?1498653708=&response-content-disposition=inline%3B+filename%3DMathematical_modeling_of_Brown_Seaweed_D.pdf&Expires=1730698473&Signature=flMCPS1F~0zADqMRidL2kT40tR8lIlibqMz4s6h6bIlsEyRNa51HtBAe3QZhpuiM0GGGR~XS3Ujb7lPXs2OjCr6DLW0I5dvqQxl-xykBYmZ0fVcr280ywxn8TUllRSkNg0dHJa20J3bkdAhDuS6LNYDfGLnugnTrZ6qFqV2NTizdXiCcqiqDTAa0QkdTailQMzqpJmElNx3l9LCnrz4ouqNvrbxGu2wJc~IAvRp0DQqLauHCygWYG2~Hq5Y6KrzYceZed9J8XCDvKNIEJN~pidOKaCkHP85WAxIQh5kfg-A-FrnnqexWF0Cb6crwvUDU6KBU1dW7qIGOJL8jFpXemw__&Key-Pair-Id=APKAJLOHF5GGSLRBV4ZA

Fudholi, A., Ruslan, M. H., Othman, M. Y., Azmi, M. S. M., Zaharim, A., & Sopian, K. (2012b). Drying of palm oil fronds in solar dryer with finned double-pass solar collectors. *WSEAS Transactions on Heat and Mass Transfer, 4*(7), 105–114. https://wseas.com/journals/hmt/2012/54-534.pdf

Fudholi, A., Sopian, K., Ruslan, M. H., Alghoul, M. A., & Sulaiman, M. Y. (2010). Review of solar dryers for agricultural and marine products. *Renewable and Sustainable Energy Reviews, 14*(1), 1–30. https://doi.org/10.1016/j.rser.2009.07.032

Ge, L., Wang Peng, W. P., & Mou HaiJin, M. H. (2011). Study on saccharification techniques of seaweed wastes for the transformation of ethanol. *Renewable Energy, 36*(1), 84–89. https://doi.org/10.1016/j.renene.2010.06.001

Gupta, S., Cox, S., & Abu-Ghannam, N. (2011). Effect of different drying temperatures on the moisture and phytochemical constituents of edible Irish brown seaweed. *LWT-Food Science and Technology, 44*(5), 1266–1272. https://doi.org/10.1016/j.lwt.2010.12.022

Han, J. W., & Keum, D. H. (2011). Thin layer drying characteristics of rapeseed (*Brassica napus* L.). *Journal of Stored Products Research, 47*(1), 32–38. https://doi.org/10.1016/j.jspr.2010.05.006

Ibrahim, M., Sopian, K., & Daud, W. R. W. (2009). Study of the drying kinetics of lemon grass. *American Journal of Applied Sciences, 6*(6), 1070. www.scipub.org/fulltext/ajas/ajas661070-1075.pdf

Jain, D., & Tiwari, G. (2003). Thermal aspects of open sun drying of various crops. *Energy, 28*, 37–54. https://doi.org/10.1016/S0360-5442(02)00084-1.

Kant, K., Shukla, A., Sharma, A., Kumar, A., & Jain, A. (2016). Thermal energy storage based solar drying systems: A review. *Innovative Food Science and Emerging Technologies, 34*, 86–99. https://doi.org/10.1016/J.IFSET.2016.01.007.

Kilic, A. (2009). Low temperature and high velocity (LTHV) application in drying: Characteristics and effects on the fish quality. *Journal of Food Engineering, 91*(1), 173–182. https://doi.org/10.1016/j.jfoodeng.2008.08.023

Meziane, S. (2011). Drying kinetics of olive pomace in a fluidized bed dryer. *Energy Conversion and Management, 52*(3), 1644–1649. https://doi.org/10.1016/j.enconman.2010.10.027

Ong, K. (1999). Solar dryers in the Asia-Pacific region. *Renewable Energy, 16*, 779–784. https://doi.org/10.1016/S0960-1481(98)00279-1.

Othman, M. Y. H., Fudholi, A., Sopian, K., Ruslan, M. H., & Yahya, M. (2012). Analisis kinetik pengeringan rumpai laut *Gracilaria changii* menggunakan sistem pengering suria. *Sains Malaysiana, 41*(2), 245–252. www.ukm.my/jsm/

Rathore, N., & Panwar, N. (2021). Solar drying. In *Fundamentals of renewable energy*. https://doi.org/10.2212/spr.2008.2.9.

Ruslan, M. H., Fudholi, A., Othman, M. Y., Azmi, M. S. M., Yahya, M., Zaharim, A., & Sopian, K. (2011, October). The double-pass solar dryer for drying palm oil fronds. In *Proceedings of the 10th WSEAS international conference on system science and simulation in engineering (ICOSSSE'11)* (Vol. 2011, pp. 143–149). Penang, Malaysia. ISBN: 978-1-61804-041-1. https://d1wqtxts1xzle7.cloudfront.net/96863519/ICOPOW-24-libre.pdf?1672931131=&response-content-disposition=inline%3B+filename%3DThe_double_pass_solar_dryer_for_drying_p.pdf&Expires=1730698566&Signature=fr8o25y2VZH8jcsWsKCTPcgsY4sFpU00al8oyY-Lp0r6bxbRWuohrYoX6g~GOyNFjKv8wGXLUBax70xyDEOrPiZ7UDcS55u-3v9MY64nMhTtCbg0m6FrFQNh3aT-GwPZdXqiT4MdBMavc2O42lINgtO5BcKaXRs5J84lUzAgNXaEDAIBP6jio1jt8m0bPkthYdOoUnybuXgVP9i9owdad~6c8MdBXKFTXvzBqciSdXOFCXQoSx0NDt8Hc8MNKlF9ISRgX~o4buYrl5TWpRjcXGqCOa5LWUFFadUErw9q6FDfNTduQl~vBeXP6ZkFnUkZT8HrSlBzf17ZdaFbwDIFfQ__&Key-Pair-Id=APKAJLOHF5GGSLRBV4ZA

Simal, S., Femenia, A., Garau, M. C., & Rosselló, C. (2005). Use of exponential, Page's and diffusional models to simulate the drying kinetics of kiwi fruit. *Journal of Food Engineering, 66*(3), 323–328. https://doi.org/10.1016/j.jfoodeng.2004.03.025

Singh, G. (2013). Solar power generation by PV (photovoltaic) technology: A review. *Energy, 53*, 1–13. https://doi.org/10.1016/J.ENERGY.2013.02.057.

Soares, O. D., da Cruz, A. M., Pereira, G. C., Soares, I. M., & Reis, A. J. (Eds.). (2007). *Innovation and technology—Strategies and policies*. Springer Science & Business Media. ISBN 0-7923-4435-9

Sopian, K., Othman, M. Y., Zaidi, S. H., & Amin, N. (2012). Advanced solar assisted drying systems for marine and agricultural products. *Journal of Mechanical Engineering, 42*(1), 9–14. http://ases.conference-service.net/resources/252/2859/pdf/SOLAR 2012

Tunde-Akintunde, T. Y. (2011). Mathematical modeling of sun and solar drying of chilli pepper. *Renewable Energy, 36*(8), 2139–2145. https://doi.org/10.1016/j.renene.2011.01.017

Vergara-Fernández, A., Vargas, G., Alarcón, N., & Velasco, A. (2008). Evaluation of marine algae as a source of biogas in a two-stage anaerobic reactor system. *Biomass and Bioenergy, 32*(4), 338–344. https://doi.org/10.1016/j.biombioe.2007.10.005

Index

O

P

Q

R

S

For Product Safety Concerns and Information please contact our EU
representative GPSR@taylorandfrancis.com
Taylor & Francis Verlag GmbH, Kaufingerstraße 24, 80331 München, Germany

www.ingramcontent.com/pod-product-compliance
Ingram Content Group UK Ltd.
Pitfield, Milton Keynes, MK11 3LW, UK
UKHW021506240926
475702UK00035B/505
* 9 7 8 1 0 3 2 6 7 8 2 3 8 *